From Margins to Medicine

A First-Generation Student's Health Equity Guide
on *Overcoming Adversity with Diversity*

WILLIAM MUNDO

Foreword by BRENDA J. ALLEN, PhD

ISBN: 978-1-7357941-0-5

Twitter: @Zapantera
www.linkedin.com/in/williammundo
https://www.linkedin.com/in/williammundo/Mundogoingglobal.blogspot.com
Williammundo.com

Cover design by 100Covers
Interior design by FormattedBooks

About the Author

★ ★ ★

William Mundo was born on March 19, 1994, to the parents of immigrants from Acapulco, Mexico. As a son of two immigrants, he is also a First-Generation New American and a First-Generation medical student. William is an older brother and the oldest cousin in his family's generation. He was born in urban East Los Angeles, California, and was raised in rural Colorado. On top of having a burning passion for medicine, he is also enthusiastic about diversity, equity, and inclusion. He has an appreciation for civic discourse, social justice, history, and politics. He enjoys spending time with his family or spending it outdoors, writing poetry, mixing music, and playing basketball. William earned a Bachelor of Science in public health and a Bachelor of Arts in ethnic studies from the University of Colorado Denver. Shortly after his undergraduate studies, he also received a master's degree from the Colorado School of Public Health. Currently, William is a medical student and well on his way to become a medical doctor at the University of Colorado School of Medicine.

To get the best experience with this book, I've found readers who read and understand *Difference Matters* by Dr. Brenda J. Allen are able to implement faster change and take the next steps needed to achieve health equity as First-Generation students and New Americans.
You can get a copy by visiting: www.differencematters.info

Dedication

★ ★ ★

Family and ancestors: you are the reason why I wrote this book. My mother – your sacrifices allowed me to live the life of my dreams. My father – who taught me my way of living. My grandmother, who raised me with the love and compassion I now serve my patients with. My little brother – for your inspiration to make me a better role model. My uncles and aunts – who would always believe in me. My cousins – you all inspire me to be a better role model. My wife who has given me new meaning in life. This book is for you.

Mentors and peers: you have been my biggest support, and I could not have done it without you. I am eternally grateful for those who supported me on my journey. You have changed my life completely. I can't wait to keep making the world a better place with you and our generations to come.

The margins: to all of those who are immigrants or children of immigrants or those who grew up in underserved communities and wanting to pursue an education, this book is for you.

Table of Contents

Part 1: The Humans of the World

Part 2: Translating Theory into Practice through Storytelling

Foreword

★ ★ ★

When I met William Mundo, I knew that he was someone special. He was an undergraduate student at the university where I was the Chief Diversity Officer. The son of undocumented immigrants from Acapulco, Mexico, and the first in his family to go to college, Will knew his life purpose. His goal? To change the world. How? By becoming a medical doctor practicing at a global scale to tackle health disparities. It's fitting if not prophetic that Will's surname translated to English is "World." Indeed, some of us refer to him as Dr. World! That was over eight years ago. Since then, I have interacted with Will countless times and I have observed or heard about his unrelenting efforts. His vision, passion, and dedication have remained clear and unwavering. He has steadfastly and successfully worked towards his goal.

Will is definitely on track to achieving his ultimate goal, and he already is making a significant difference. He has earned accolades and awards while serving in a wide variety of roles. These include being a leader of student groups, a peer mentor, a social justice advocate, a volunteer trainer (for community health workers in Peru), a fellow in an undergraduate pre-health program, the co-founder of a Latinx Medical Student Association, and many more. He engaged in these and other endeavors while completing undergraduate and graduate degrees, as a medical student, and being employed. Will's tireless willingness to serve and to lead demonstrate his enduring commitment to help, to give back, and to make a difference for underserved, underrepresented groups. *From Margins to Medicine* is destined to become another way that he will impact the world.

Will aptly depicts the context for his book: "Our world is experiencing an unprecedented time with the COVID-19 pandemic, the accumulation of several hundred years of socioeconomic and racial oppression, and a renewed and unfinished civil rights movement. The historical authoritarian underdevelopment of nations has shown that living conditions for people worldwide are not equitable. Our own country is wounded and injured by the historical traumas we have endured over several generations." Will offers a wealth of insights and information to guide readers to address these and related challenges. Written in a format that Will describes as a "memoir-style reference book," *From Margins to Medicine* combines storytelling with explanations and examples of theories and concepts from various areas of studies, including health sciences, education, diversity, and social justice.

Will speaks specifically to First Generation Americans (FGA), those who are the first in their family to navigate various, interconnected institutions in American society, the first in their family's generation to either be born in the United States and/or to immigrate to the US and pursue an American education. He notes, however, that he does not speak for all FGAs. Rather, he stresses similarities of struggles for individuals who grew up on the margins of U.S. society. And, he provides new perspectives on how we can move together as a nation of immigrants to heal and be healthy.

Please note that Will's intended audience expands beyond FGAs. As he explains, "Whether you are a First-Generation American, an ally of health equity, or a person who needs a new framing of knowledge and awareness or is just curious to learn—this book is for you."

"Part I: The Humans of the World," provides a foundation and framework for the book. Six chapters contextualize and introduce fundamental concepts of health equity, including race, social determinants of health, power and wellness, racial health disparities, and impacts of generations of oppression, including systemic and individual pathologies. Will narrates lessons learned from public health and medicine as a First-Generation American, and as the first in his family to navigate various institutions. He vividly describes his unique challenges and connects them to impacts on public health. Part I concludes with an alternative perspective on immigration history based on perspectives of descendants of Acapulco, Guerrero, Mexico.

"Part II: Translating Theory into Practice through Storytelling," elaborates on concepts from Part I. Will clarifies and exemplifies practices with concrete, real-life examples, including personal life case studies. He critically analyzes complexities of being a first-generation student pursuing the American Dream. He acknowledges obstacles while also describing how to anticipate and overcome them. For instance, the story of his beloved grandmother's health challenges vividly illuminates how the social determinants of health are charted throughout one's life and across multiple generations. In addition, his stories reveal implications of structural and systemic racism in institutions such the criminal justice system and education. He narrates stories of resistance and resilience, including his own persistence to overcome tragedy and inequities in medical, education, political, and criminal justice systems. Part II details a call to action to transform inequities in the United States and the world. It includes practical steps for how to adopt a multidisciplinary approach to addressing health inequities. The book concludes by delineating types of structural pathologies that FGAs and others might address.

From Margins to Medicine is a message and map of hope for what Will describes as "breaking an intergenerational structural and cyclical pathology that affects our human condition." Regardless of what motivated you to read this book, prepare to be educated, empowered, and equipped to confront and conquer global healthcare challenges.

Due to his background, his knowledge, his compassion, his contributions, his wisdom, and his experiences, Will is uniquely qualified to be the author of this timely, powerful book. I am grateful that he invited me to write the foreword because it gave me an opportunity to learn more about him and to learn from him. As he did when we first met, Will continues to impress and inspire me. In closing, I wholeheartedly agree with Will's assertion that "Amid a pandemic that has shocked the whole world to its core, corruption in politics, and being on the verge of societal catastrophe, this book is imperative to read."

Brenda J. Allen, Ph.D.
Professor Emerita - University of Colorado Denver
Vice Chancellor for Diversity and Inclusion (retired)
Author of *Difference Matters: Communicating Social Identity*
(Waveland Press)
January, 2021

Introduction

★ ★ ★

From the first generation to the next, I became what I am today, a New American when an impossible journey to pursue the "American Dream" began in 1994 - the year I was born in America to immigrant parents. If I work hard, I can pull myself up from my bootstraps and achieve freedom, liberty, and justice. Well, wait, what if I don't have any boots? The American Dream is a cultural myth claiming to represent equal opportunity, but the truth is it just doesn't allow for the same outcome for everyone. The expectation is that no matter your background, you, too, are promised the ability to pursue prosperity and happiness. To some, no matter where you come from, and from whatever part of the world, you can rise from the margins of society and become a successful American. Yet, traditional patterns in history show that trauma across generations continues to persist, and upward mobility in America is not the same for everyone due to the unique and complicated circumstances each person faces. Certain people, especially those coming from the margins such as people with minority or poor backgrounds find themselves at a disadvantage from the get-go and result in inequities. The inequities in America include differences in health, freedom, and socioeconomic outcomes that all stem from structural pathologies that this book will discuss.

Structural pathologies refer to the diseases caused by systemic issues in society, rather than in biology. Structural pathologies are similar to structural violence, which is a term described by liberation theologians during the 1960s which describes, "social structures – economic, political, legal, religious, and cultural – that stop individuals, groups, and societies from reaching their full potential in health."[1] Death, suffering, and pain are all symptoms of a culture where socially constructed categories are the main drivers of health. Although our genetic codes are similar, our zip codes are not, and depending on where you live and grow up, it can change your trajectory in life and even determine your life expectancy. Rising from the margins of civilization to becoming a first-generation physician, I was able to witness the micro and macro interactions between the individual and the environment. To appreciate I used my diversity to overcome adversity, I had to use an interdisciplinary comprehension of human behavior and science, as well as have an open mind. With this memoir-style reference book, I will explore the relationships between race, health, and justice while I also highlight how some of these topics have intersected with my life.

My book is composed of two parts, all of which demonstrate storytelling elements combined with theory, practice, and a vision for the future for the next generation of First-Generation Americans. I am writing this book because I believe I have ideas and stories to share that can ignite change. My personal account and experiences as a first-generation American are the reason why I am writing this book. I want to show readers some of the issues that people who live in society's margins face and provide a new perspective on how we can move together as a nation of immigrants.

Book Outline

★ ★ ★

In "Part I: The Humans of the World," I will set up the context of the book and introduce fundamental concepts of health equity through an integrated approach. This knowledge is a prerequisite to innovatively create solutions for some of the most pressing issues we face in modern history while developing acculturating skills instead of surrendering to American assimilation. In chapter one, I will give an overview of the lessons learned from public health and medicine as a First-Generation American. I've encountered unique challenges and experiences that resulted in my professional and scholarly pursuits, which I will highlight in chapter two, where I introduce the history and concept of race in America and how it impacts public health. In chapter three, I will introduce the concept of the Social Determinants of Health, which are integral to understanding how structural pathologies work. Chapter four will interrogate the relationship between power and wellness and explain how it creates inequity in our society. Chapter five will demonstrate different examples of racial health disparities in America and discuss how generations of oppression can create systemic and individual pathologies. Chapter six discusses immigration history from a different historical perspective on and my view as a Native Inmigrant. Coming from the descendants of Acapulco, Guerrero, Mexico, I have been the first in my family to navigate the institutions in post-colonial America.

In "Part II: Translating Theory into Practice through Storytelling," I emphasize the concepts I talked about in part one and put them into practice with concrete, real-life examples, and personal life case studies. I critically analyze the subtle complexities of being a first-generation student pursuing the American Dream while also describing my life stories that highlight several of the ideas brought up in part one. In chapter seven, I use the story of someone very special to me and show how the social determinants of health are charted throughout life and across multiple generations. Chapter eight highlights the story of how I became inspired and share the tragic story that changed my life. Following my college journey, I will go back in time and highlight the school-to-prison pipeline in our country in chapter nine and talk about real life experience engaging the criminal justice system as a teenager. I will share my story of how I progressed through America's medical, education, political, and criminal justice system. Chapter ten described the formation of a new identity of the New American and serves as a call to action for the New American to respond to and transform inequities in our country and the world. Chapter eleven lays down practical steps for New Americans to adopt a multidisciplinary approach to addressing health inequities. Finally, chapter twelve highlights the different types of structural pathologies in society that we can address as the next generation.

Experiences and Expertise

★ ★ ★

In several chapters, I demonstrate specific moments in my life where I have overcome adversities that are commonly experienced by many people who share similar identities as me. In my life, I've used these moments as an opportunity to learn and become a better leader in health equity. On top of being a well-respected leader in the industry, I have built several alliances worldwide, some of which include forming sustainable partnerships with medical students in the Peruvian Amazon jungle, high-altitude maternal and infant researchers in La Paz, Bolivia, and farmworkers in the highlands of Guatemala. I have visited two continents including fifteen out of the 195 countries in our world.

The extensive background and training I have equipped me with the knowledge and expertise to write this book. I am a published author in peer-reviewed scientific journals. I do peer-review editing, and I have contributed scholarship to the scientific community to advance science and human rights during my early professional career. I graduated from the University of Colorado Denver with a bachelor of science in public health and a bachelor of arts in ethnic studies, and a master's degree in public health with a concentration in global health systems management policy. I am currently a medical student pursuing a medical doctoral degree. Shortly, I will begin my medical training as a physician graduate emergency medicine resident.

We all have an opportunity to learn and grow, whether in the classroom or in life. Because of these opportunities, I have committed my life to be a student, an educator, and a proponent of my beliefs, while also maintaining my integrity and inalienable right as a descendant of the Mexican people of Aztlan, which I will further discuss in the book. My preparation inside and outside the classroom has shaped my perspective, allowing me to provide the insight to help first-generation students overcome the adversities they'll face while pursuing an education in the US.

What Is a First-Generation American (FGA)?

★ ★ ★

There is currently no widely accepted term of what it means to be a First-Generation American (FGA). To me, an FGA means to be the first in your family to navigate the institutions in American society. In the broadest sense, to be an FGA implies that you are the first in your family's generation to either be born in the United States and/or to immigrate to the US and pursue an American education.[2] The central themes rely on integrating into the US and becoming American while assimilating or acculturating to its institutions. The difference between assimilating and acculturating is that you decide to let go of your culture to adopt a new one, while acculturing means maintaining your own culture, keeping it as you embrace American values. The main factor I focus on in my book to define an FGA is being the first in your family to pursue higher education.[2] Still, it can also include navigating US politics, health, and business systems.

In my case, I am an FGA for two reasons: because I was the first in my family to become a US citizen, and also because I was the first in my family to graduate from high school and pursue higher education. People begin to identify as FGAs in the educational system, primarily when nobody in their previous family generations has pursued higher education in the US. Being the first means not having generational and institutional knowledge on picking a college, applying to scholarships and financial aid, finding your calling and purpose, and transforming your life to improve the next generation of your family. As an FGA, I have been the first in my family to pursue what is referred to as the American Dream.

When I think of our country, I think of a multiethnic and interdependent collective nation. Although currently our nation is polarized, I still believe we all have a common thing that unifies us – and that is that we are all Americans. However, many have to overcome systemic challenges to achieve this American Dream. The American Dream is supposed to be a promise and an equal opportunity to contribute to humankind's advancement. Still, we never discuss how we can become more equitable with distributing resources to allow equitable improvement for everyone, especially in the context of historical injustices. Whether you're talking about the Indigenous holocaust of America, the African-diaspora slave trade, the exclusion or internment of Asian ethnic groups, or the mass incarceration of Black and Brown Americans, all historical events are rooted in structural pathologies of power and the belief that one race is superior to another. As incoming FGAs, we must learn how to use our diversity to create equity and achieve our own *New American Dream*.

The Benefits of this Book and the Next Steps for the New American

★ ★ ★

In this book, I share with you what we need to learn to become the next generation that is dedicated to justice, diversity, inclusion, and equity for all of human life—not only for us in the US, but also for everyone and their future generations who live all around the world. As New Americans, it is our moral obligation to do no harm to individuals and communities and build alliances across the globe and use the knowledge of emerging science and our ancestors to allow us to transcend the different worlds of medicine and justice. The main lesson of this book is to discover the connections between health and ethnic scholarship and provide inspiration for the next generation of FGAs who will change our future. If you are an FGA, this book will help you comprehend how you can come from a diverse background and, despite adversity, still pursue a New American Dream.

The benefits of reading this book include learning to advocate for the things you love or believe in, converse about specific issues with a framework, develop educated opinions, and understand the latest research in health equity. At the end of this book, you will obtain the perspectives that will allow you to thrive in any professional career and in your own personal life as an FGA or a concerned ally that wants to improve our society. I do not speak for all FGAs, but I know that there are many of us from various backgrounds who share the same experiences and struggles as a consequence of growing up in America's margins. I also acknowledge that there will be experiences I can't comprehend because they differ from my own due to my own privilege. Even though this is true, my humility and self-awareness have allowed me to acknowledge others' existence and life experiences.

I use an educational and inspirational pedagogy combined with memoir-style anecdotes throughout the book. I will discuss several counternarratives that run contrary to the mainstream thinking we are taught to believe in the traditional American thought. Using a scientific framework, I will provide several opportunities for you to learn more about overcoming the adversities of American society and how to become a successful FGA. We must be clear in our understanding that not everyone begins with equitable opportunities, and as a result, the chances of making the American Dream come true are more challenging and sometimes unfeasible. However, there are still possibilities for all of us to learn how to become better. I hope that through my story, you will appreciate that even if we live on the margins of American society, there's hope in breaking an intergenerational structural pathologies that impact our human condition.

Amid a pandemic that has shocked the whole world to its core, corruption in politics, and being on the verge of societal catastrophe, this book is imperative to read. Our world is experiencing an unprecedented time with the COVID-19 pandemic, the accumulation of several hundred years of socioeconomic and racial oppression, and a renewed and unfinished civil rights movement. The historical authoritarian underdevelopment of nations has shown that living conditions for people worldwide are not equitable and magnifying the inequities that exists in our country. Our own country is wounded and injured by the historical traumas we have endured over several generations. Learning about these traumas during my upbringing and in my educational career, I have learned how to address them and how to begin to heal from them.

If you are an FGA student aspiring to become a physician, this book is for you. If you are an FGA teenager who lives in a rural or urban underserved area and feels trapped in the system, this book is for you. If you are an ally wanting to learn about the cultural experiences and perspectives of others, this book is for you. If you are a student of wellness, this book is for you. In my life, I have learned to relish and tell my own story, which has taught me to be a better and kinder human being. In the same way, I hope learning about my story in this book will introduce you to do the same thing to change the track of human history and leave your mark as an FGA.

You may feel uncomfortable reading this book, but you will benefit the most from entering it with an open mind. You might feel uncomfortable because you will learn about topics that aren't often discussed in traditional spaces in education. I want to remind you that it's okay to feel uncomfortable; it is part of the growing process. Whether you are a FGA, an ally of health equity, or a person who needs a new framing of knowledge and awareness or is just curious to learn—this book is for you.

REFERENCES

1. Farmer PE, Nizeye B, Stulac S, Keshavjee S. Structural violence and clinical medicine. *PLoS Med.* 2006;3(10):e449. doi:10.1371/journal.pmed.0030449
2. Jehangir RR. Higher Education and First-Generation Students: Cultivating Community, Voice, and Place for the New Majority. 1st ed. New York: Palgrave Macmillan; 2010.

PART 1

★ ★ ★

The Humans of
the World

First-Generation Experiences in the United States

Overview of the lessons learned from public health and Medicine as a First-Generation student

"Health equity means that everyone has a fair and just opportunity to be healthier. This requires removing obstacles to health such as poverty, discrimination, and their consequences, including powerlessness and lack of access to good job with fair pay, quality education and housing, safe environments, and health care."

—Robert Wood Johnson Foundation, Definition of Health Equity

Learning the Connection Between Opportunities and Structural Racism

★ ★ ★

When I was a freshly minted high-school student, I remember sitting in our small school's library, meeting with students who had recently graduated from high school and pursuing a college degree. I was bored, not interested, and wanted to leave already to hang out with my friends. I would always sit with my friends in the back of the library just so we could mess around. When one of the speakers asked the audience, who was interested in going to college, several hands went up. But as I looked at my friends, we all didn't raise our hands. I knew that college was a word that meant nothing to us. It was talked about among some of my older peers, my white friends, and in some classrooms; however, we were afraid to ask what it entailed, especially since it seemed like everyone else knew what it meant. Besides, we all knew that people like us did not go to college.

As the conversation in the library continued, the presenters described college as, "more school." When we heard that we were certain that we were not meant to go to college because who would want to do more school? School was hard for us already as it is. During the discussion, someone asked about getting books, and the speaker answered that you could just buy them online or borrow them. "Buy books?" I yelled, and nervously laughed. I was puzzled. Why would anyone want to spend money on buying books? At this time, I started to pay more attention to the discussion and was curious to learn more as to why someone would do something so bizarre to buy books for school. I thought the books were supposed to be provided by the school, or so I thought. The only book I would even consider ever buying was maybe a science book, because I know that is where my passion was and what I found fascinating. I later discovered that science is the foundation of medicine. At the end of the discussion, I learned that I had to go to college if I wanted to become a doctor and that there was no other way around it.

When I decided to share with people that I wanted to become a doctor, many laughed, and others discouraged me. Someone said to me, "Mexicans don't become doctors; they become laborers for construction instead." Or "Why don't you just become a mechanic?" I believed this to the point that I began working in construction at the age of fourteen. I remember having to shovel several hundred pounds of snow and carrying massive rocks over 50 lbs up and down four to five sets of stairs. To be honest, I was miserable, and I was able to see that the work was taking a toll on the elders' health, several of who I worked with. By no means am I trying to throw shade to those who work in construction or mechanics, I believe they are also integral to the functioning of our society. However, I just knew that I did not want to follow this path for myself. Fortunately, I had some teachers who believed in me and my ability to pursue my educational goals. In these moments, I learned that instead of following the footsteps that I had been told to follow, I could just create my own path instead. That is the gift of being a first-generation student (FGA)—you create your own path.

No matter what side of the aisle you represent or where you are located in the world, we can all agree that we all deserve an opportunity to pursue a quality education. We must stop structural racism in education as it is a form of denying knowledge to people who traditionally do not have access to education. Structural, institutional, and systemic racism broadly refers to the structures set in place to disadvantage minorities which include rules, practices, customs once rooted in law now with residual effects that permeate throughout all aspects of American society.[1] Some examples of different forms of structural racism includes discriminatory hiring and advancement practices, denial of access to resources and wealth, and barriers set in place to prevent racial minorities from succeeding in education.

One particular example that caught my attention was with the journey of Malcolm X, who was denied an opportunity to education and instead was forced through the school-to-prison pipeline. He experienced the same systemic inequities all FGAs do and did not pursue a formal education; however, he continued to articulate his aspirations for America's potential and the issues we faced as a country. Of course, people who write traditional history books misinterpret several narratives oftentimes and only write one sided stories. Some will say Malcolm was just a "radical negro," but they know little about his life and what he had to go through. Malcolm aspired to become an attorney, but that opportunity was denied to him because of his skin color. Instead, he was relegated to become a carpenter. Under the veil of racism, Malcolm was robbed of an opportunity to earn a law degree and instead pursued an education while in prison. Even though he had no formal education, he received first-hand knowledge of the criminal justice system when he served a prison sentence for several years. Shortly after that, he became a resounding voice in the political system as a prominent leader in civil and human rights. Malcolm's story is just one example of why we should continue affirmative action programs beyond educational systems to restore justice.

Malcolm's story is not a unique one, and one that obviously tends to still happen even in modern history. Many people I know are still not able to pursue an education for the same exact reasons. The racial oppression that occurs at an individual and systemic level will be a battle that we will continue to face. Being an FGA, it is crucial to be aware of our own racial and ethnic identities and histories and encourage others and ourselves to become anti-racist. We must understand racism as a socioecological pathology—a result of society and economics, not biology. We must unlearn myths and falsehoods as we ask ourselves about the truth of the human condition.

This book is the collective narrative of those who have created so much positive social change under more meager and fatal social conditions. As FGAs, we should be inspired to make a difference in other people's lives by improving access to opportunities for our families, patients, and communities—and that includes access to not just education but also health care. We have a moral obligation to instill a sense of empowerment and self-efficacy in those who look up to us and challenge those who look down on us. My hope is that the lessons in these pages will guide the next generation of FGAs, just as those warriors before us who have improved our lives and have gave us an opportunity to create our own paths.

One of the most important lessons I've learned from becoming an FGA is to build reciprocal relationships with family, friends, colleagues, and the community. There is no doubt that we must be an active and invested participant in our lives and those around us. We can attain this by developing our own leadership skills, understanding the vital intersections of power and health in the context of engaging in service learning.

Service Learning in Communities

★ ★ ★

The opportunity to go to college gave me exposure to learn about the importance of service learning and also how to best engage in service learning with communities. Service learning refers to a, "structured learning experience that combines community service with preparation and reflection with the goal to provide service in response to community-identified concerns while making connections in academic coursework and our role as citizens of this world."[2] In 2012, I was admitted to the University of Colorado Denver with scholarship support. I was able to pursue my undergraduate studies in areas that have taught me how to serve my community. During my time at school, which had one of the most diverse campuses in Colorado, I provided services to the university community, which taught me some of the most valuable lessons I have learned in my life as an FGA. I provided direct care and services to multicultural, first-generation, low-income, and marginalized students, managing and mentoring several hundred students who needed guidance through the TRiO Student Support Services Office.

Before becoming a peer mentor with TRiO, I was first a participant and received life changing support that kept me engaged in college and helped me graduate. Upward Bound and TRiO SSS are federally funded programs that are designed to support students like me. The Federal TRiO programs award grants to institutions of higher education to serve students who come from underrepresented backgrounds with the overall goal to increase college retention and graduation rates of participants.[3] Through TRiO, some of the services included providing one-on-one mentoring, group mentoring, tutoring, and also providing workshops in professional and educational development. With the help of this program not only was I able to graduate from college, but also learn more about my professional and academic interests.

During this time, I learned how vital my local and the broader community are in improving equity for everyone and how to collaborate well with others. When working with these students, I emphasized the role of service learning and how it is different than community service or volunteering. The main difference is that service learning has a bigger purpose than just doing something to add to your resume, to fulfill probation, or just because you need to kill time. Service learning is an organized method to ensure that you meet the actual needs of the community you want to serve.

Rather than approaching service using a deficit model, that is only looking at the things that need to improve, I learned that you must leverage community assets and strengths simultaneously. By deficit model, I mean that we tend to only look at the bad aspects of a community and what it is missing rather than appreciating the things that are working well. One example of how people use a deficit model while serving a community is when we focus on what's missing. The most common example is thinking about a community that does not have any shoes. Seeing this as something that is missing, some well-intended people might decide to get shoes for the whole community. However, once they get access to those shoes they won't use them because they don't use or want shoes. Another example is the lack of hiking trails or biking paths in the neighborhood. Instead of creating these paths, perhaps community members just like to use the park down the street for recreation. With this change in mindset and perspective, we can focus our attention on developing the park to make it more accessible instead of using resources on something that the community will probably never use.

The need to serve communities' needs and make a difference in people's lives will always be a primary motivation for FGAs who want to make our country a better place. When you give back to the community, you are not helping or fixing but serving with a purpose. It is vital to use education, direct service, and reflection to give back. There is a need to act alongside people who are like us and those who are different from us, including those from the communities themselves—only then will be able to create meaningful change. There must be a targeted and nuanced approach to the way we look at service and practice health care to meet the exact needs of the communities we serve.

Another opportunity I received from my service learning was to witness climate change's impact on the communities I served on the ground across the world. Whether working alongside flooded river communities in the Peruvian Amazon Jungle, working with the Mayans' descendants in the droughts in the highlands of Guatemala, and cities wrecked by earthquakes alongside the coast of Ecuador, I have seen first-hand impacts of climate change. Although we are some of the largest contributors to climate change here in the US, we are also the least likely to face the consequences of our actions on the climate.[4] This becomes important because it creates more inequities in our world. It is unethical that the largest contributors to climate change are also the ones who are least impacted by it and do nothing to change that fact.

Historically, pollution continues to get worse and has many adverse effects on human health. With industrialization, human production rates of greenhouse gases are at dangerous and unprecedented concentrations. These disasters all bring about social vulnerability and affect the health of people depending on region and population. Areas that are at risk from combined climate impacts will see concurrent environmental or socioeconomic stresses in their communities that will affect their overall health. We must move toward practices and policies that help mitigate the effects of human behavior on our climate, as this is already the greatest threat to human existence on our planet.

Although climate change is one of the main factors that negatively impact underserved communities' communal health, there are many other factors I have personally seen and studied that are also detrimental to the health of our communities. For example, racism has mental health consequences for many people because of the internalization of ongoing person-to-person discriminatory interactions and discriminatory or inadequate policies, such as providing insufficient funding to ensure mental health services for people in the criminal justice system.[5] Another factor is food insecurity and the lack of access to fresh produce to ensure healthy meals for everyone instead of a gas station and fast-food options, which are all typically cheaper.[5] Although initially more affordable, the long-term consequences fast food has on our health is expensive and deadly. Decreased access to public services, transportation, education, and representation are all other factors that negatively impact underserved communities' health.[5] Service learning has allowed me to study these issues and work alongside people from all walks of life and ensure a better future for all.

Diversity in Health Care and Community Health

★ ★ ★

Diversity in health care and community health is imperative to overcoming many of the barriers to achieve better health outcomes for all. The concept of diversity includes the range of human differences and elements, including but not limited to race, ethnicity, gender, sexual orientation, age, and social class. Diversity has taught me that most health care happens outside of the clinic or hospital. I have learned about many direct and indirect factors that have a significant contribution to the community's cumulative health. Several of these factors include social representation in politics, infrastructural developments, legal protections, policies, and wealth. Everything is interconnected, and a shortcoming somewhere has the potential to cause deficiencies in other areas. In other words, we coexist with others, sustaining an equilibrium.

Outside the healthcare system, we, in fact, have the most tremendous potential to address health-related issues. It's a paradox because we often think we need to see the doctors when we are sick, which I believe is true to some capacity. However, this is only part of the equation for what it takes to reach equitable health outcomes. Once someone goes to the doctor, it is usually already late in the disease, which generally means that the health issue has already manifested (not counting preventive and annual visits). The work that can be done outside of the hospital or clinic and in other areas in the community is meant to target the potential root cause of health issues and eliminate them before they can even manifest themselves in our health care institutions. Even if we just rely on our healthcare system to be efficient, it will not be sustainable and effective. This is mostly because the US healthcare system is not diverse and is ill-equipped to grapple with health care issues that need to be addressed at the communal level, rather than the individual.

Although shortcomings in the healthcare system have significantly failed many people, I've learned to appreciate that failure is not always a negative thing. Instead, it is what we do with the opportunity to grow. Failure opens the door for the opportunity for patients and providers to help each other through mutual understanding and building a therapeutic alliance. Learning how to fail is a valuable experience we must learn as FGAs and healthcare professionals. We must continuously display patience and be keen to overcome the limitations of medicine, science, and human behavior.

I have learned that diversity and inclusion, and health equity have enormous benefits in science and health care for several reasons. Some of these benefits include having a robust workforce that can meet the needs of our changing demographics. As urbanization continues to grow every day and populations expand rapidly, we need effective ways to respond to the demand. The primary way to address demand is to provide a culturally diverse workforce that can communicate with the communities they serve. A culturally diverse workforce ensures that various people know the individual, unique, and different concerns that an organization may have. This approach allows us to create a force that represents a large community more accurately and works collaboratively to pinpoint problems, discover solutions, and layout practical next steps. On the contrary, an ineffective way would be by denying preventable health services, denying immigrants from contributing to American society, and ignoring the problems we face as a nation, including racism, climate change, and the safety of our children and their children.

There are uniquely passionate FGAs in health care who have made substantial, lifesaving contributions to improve the American people's health. For example, Mexican American Dr. Alfredo Quinones-Hinojosa was an undocumented migrant farmworker who is now saving lives at Johns Hopkins as a neurosurgeon, operating on about 250 brain tumors every year and conducting research that helps fight brain cancer. Colombian American Dr. Juan Carlos Caicedo is an adult and pediatric transplant surgeon who has helped develop the country's first Hispanic transplant program. Lastly, Peruvian American Dr. Roberto Flores is a world-renowned plastic surgeon who invented the Smile Train Virtual Surgery Simulator, the most widely used surgical simulator for cleft lip palate reconstruction. These are just a few real-life examples of how diversity contributes to improving lives in a way that homogeneity would not. What do they all have in common? They are all diverse, innovative, hardworking people, qualities that define us as Americans.

Many of the lifesaving inventions and services provided by FGAs are a public benefit that everyone can enjoy. Providing care to some of the most vulnerable populations and creating programs and interventions that reduce disparities save money. In a study that examined if reducing racial-ethnic mental health disparities in mental health care can offset costs of care, it was found that by eliminating medical health disparities for Blacks and Latinos, our nation could save up to $1 billion in inpatient general medicine expenditures.[6] While these savings highlight a critical need for diverse immigrant providers familiar with the community's needs, the principal benefit is that we could drastically reduce racial health disparities in our country. A diverse immigrant health workforce is crucial to eliminate health inequities, a service that will not only improve the lives of our nation's most vulnerable but all Americans.

Experiencing first-hand the diversity across health professions has allowed me to widen the scope of my understanding of the different roles that FGAs have in society. In health care, I have the privilege of working with many other people because everyone gets sick. I see people with various conditions that I study in books, and I learn their stories, which has helped me better understand how diversity benefits not just health care but also other industries. I have learned that my role is essential, but different roles are equally important in moving our society forward. These roles include being the next generation of doctors, scientists, lawyers, politicians, teachers, and leaders, to name a few. Solving the health inequities, we face today will require more than just the health care field.

Culturally Competent vs. Culturally Responsive

★ ★ ★

Another lesson I have learned is that to achieve the best outcomes possible, we must aim to be as culturally responsive as we can. Cultural responsiveness, like the term "cultural competence," promotes and understanding of culture, ethnicity, and language. The difference between the two is that "responsiveness," does not imply that one can be perfect and have attained all the skills and views needed to work with culturally diverse people. Note, I say responsive and not competent for several reasons. The first is that cultural responsiveness operates under the premise that culture is fluid and always changing. As such, we must be well equipped to respond to our ever-changing cultural environment. Further, it also means that you can't become "competent" in someone's culture. For example, I wouldn't consider myself culturally competent even in my own culture because it assumes that I know enough and everything I need to know. Even if someone went and got a Ph.D. in Mexican American studies, it would be difficult to say you are competent in a culture that is changing every year. Culturally, competency is simply not achievable, so we must instead strive to be culturally responsive.

We must continuously seek opportunities to work with different people and have different life experiences. We can become more culturally responsive by doing our own research before going into a community we will be working with. Instead of just showing up, make sure we are informed of the things the community has been through and the things that matter to them.

The best way to increase cultural responsiveness is by diversifying your peers and experiences or traveling to different parts of the world. However, I understand that this takes a certain level of wealth and privilege, and I realize it's not realistic for everyone. Considering the impact that COVID-19 has had on our world, traveling has not been available for many people, especially those who cannot afford it or who can't leave the country. The great thing about all of this is that we don't have to fly across the world to engage with new cultures with modern technological advances. Sometimes, you can even start doing it in your own neighborhood. For example, there is a park by my house that serves a large population of Somalian refugees. When I go running or walking my dog, I often see them playing basketball, hanging out on the grass, or just walking. There, I have had several opportunities to learn about people who may be different from me. To be culturally responsive means that we are committed to becoming lifelong learners and seeking diverse experiences.

The most valuable thing I have learned is that we can address the health issues of communities that traditionally do not trust the health care system by being culturally responsive. Health is one of the most personal possessions anyone has. It takes a lot to trust another individual with your life and health altogether. It is challenging to develop these relationships with people who continuously demonstrate hate toward your existence. To fully reach optimal health, we have to build relationships with each other while understanding what has happened in our history. We must discover what some people's best interests are and seek opportunities to build interdependence instead of falling back on antiquated practices of racism, colonialism, and discrimination.

Discovering the Tribulations of Medicine as a Minority

★ ★ ★

One of my most essential experiences along this journey was learning how difficult it was to become a physician in America, all the way from how expensive it is to how time consuming it is. It feels like there are barriers left and right, even in the admission process, for example, not having any idea of how to study for your admission tests, not knowing the difference between primary and secondary applications. While going through the steps of what it took to get to the position I am currently in, I learned to see and address several of the issues that have long persisted in the American educational, political, and medical system.

Some of these medical issues are rooted in scientific racism, which I will elaborate on in chapter two. By learning more about the science of medicine, I also learned about medicine's history, such as the long, traumatic history of how medicine has evolved and continues to impact ethnic minorities. An ethnic minority is a group of people who differ in race, color, national, religious, or cultural origin from the dominant group – often the majority of population of the country in which they live. Ethnic minorities have been repeatedly used in medical experiments without their consent, and the industry took that information to advance humanity in the name of science. The most notorious example is the Tuskegee Syphilis Study, which denied medical treatment to Black Americans who had syphilis to study the disease's natural course.[7] Another example is the case of Henrietta Lacks and how her cells were stolen from her and exploited, which we will examine in detail in part two. From surgeons practicing procedures on slaves to modern-day research investigating HIV/AIDS in vulnerable populations in countries in Africa and over Latin America, these are all clear unethical examples of how scientific racism has persisted in the US.[8]

Countless ethnic minority groups were used in these studies in immoral ways to further medical research that has contributed to modern medical advancements. And the medical community should not only feel ashamed of these atrocities against humanity. Still, it should also always prioritize treating and protecting the historically medically underserved communities that have been exploited. With knowledge of these moments in the history of medicine, we now understand the unique role and responsibility that the scientific community has for advancing humanity and protecting it from social harms.

I have learned that the generational experiences in oppressed ethnic minorities have created negative perceptions of the community's health system. Ethnic minorities have a cognitive dissonance with medical providers—because they have experienced a lifetime and generations of unjust and unequal treatment by people who are supposed to heal them and cause no harm. There is an imbalance of access and, most important, healthcare justice. Overall, the main thing I learned is that these experiences have opened an avenue of unanswered questions of how to move forward with modern research and better understand historical trauma that impacts health outcomes to this day.

These experiences combined have shown me that we all have the potential to revolutionize the way we look at the educational, political, and social system in American society. Our public health equity movement relies on education and politics for critical consciousness because it allows people to take a stand, reframe their understanding, question the system, and begin a genuine transformation that can break down systemic issues of marginalization. True freedom for all people includes reclaiming Native sovereignty, Black liberation, Asian pride, and white decolonization. Only then will we lead to all people's true liberation because we aren't free until we are all free.

Because of the many lessons, I have learned as an FGA, I firmly believe that giving back to the community is one of the most vital contributions we can make. To become a successful First-Generation American, we must realize that we will be working in communities that we might not understand. Instead of feeling threatened or discomfort by difference, we must learn to embrace them. After examining my experiences and background, I know that I am on the correct path toward becoming a medical doctor. I love what my passion calls me to do, and I hope this book will show you a similar track. There is no better feeling than giving power to the people, and my hope is to show you that some of the power you have will be unlocked in this book.

REFERENCES

1. Bailey ZD, Feldman JM, Bassett MT. How Structural Racism Works - Racist Policies as a Root Cause of U.S. Racial Health Inequities. N Engl J Med. 2020 Dec 16. doi: 10.1056/NEJMms2025396. Epub ahead of print. PMID: 33326717.

2. Butin DW. *Service-Learning in Theory and Practice: The Future of Community Engagement in Higher Education.* 1st ed. New York: Palgrave Macmillan; 2010.

3. Education USDo. Federal TRIO Programs. Economic Opportunity Act. 1964.

4. Rahm D. Climate change policy in the United States: the science, the politics, and the prospects for change. Jefferson, N.C: McFarland & Co; 2010.

5. Barr DA. Health disparities in the United States: social class, race, ethnicity, and the social determinants of health. Third ed. Baltimore: Johns Hopkins University Press; 2019.

6. Cook BL, Liu Z, Lessios AS, Loder S, McGuire T. The Costs and Benefits of Reducing Racial-Ethnic Disparities in Mental Health Care. Psychiatric Services. 2015;66(4):389-396.

7. Thomas SB, Quinn SC. The Tuskegee Syphilis Study, 1932 to 1972: implications for HIV education and AIDS risk education programs in the black community. American journal of public health (1971). 1991;81(11):1498-1505.

8. Washington HA. Medical apartheid : the dark history of medical experimentation on Black Americans from colonial times to the present. 1st ed.. ed. New York: New York : Doubleday; 2006.

Biopsychosocial Origins of Race

Introduction to critical race theory in medicine and health disparities

> *"I hate racial discrimination most intensely and all its manifestations. I have fought all my life; I fight now and will do so until the end of my days. Even although I now happen to be tried by one, whose opinion I hold high esteem, I detest most violently the set-up that surrounds me here."*
>
> —*Nelson Mandela*

Race, Ethnicity and Biology

★ ★ ★

Early in my education, I knew I was Mexican not because of my skin color but because of how I was treated by authority figures. However, it all changed when I started to realize that some teachers would treat me differently. One vivid example I can remember was when I was speaking in Spanish with one of my friends who recently immigrated from Mexico. As the teacher was giving instructions, I noticed my friend was confused as to what was going on since he did not speak any English. I decided to interpret for him so he can know what was going on and what we needed to do. As I started to explain the plan, the teacher stopped us and yelled at me for speaking "Mexican." I told her that I was trying to tell him what we were doing since he did not understand, however the teacher instead insisted that I was just saying bad things about her to my friend, when in reality this was far from the truth. I tried to explain, but she quickly interrupted and said, "If you say one more word, I will send you to detention!" I got upset and kept telling her that I was not saying anything bad and that he needed help, but she just ended up sending me to detention. As I was walking away, she said, "This is America, you need to speak English only." From this day on, I knew that my Mexican identity and Spanish language was a target of racism. I just didn't know that it was racism until I was much older and better understood my racial identity. The reason why I share this story is because the idea of race and ethnicity is something that is learned early in childhood and as such is a crucial point in time where we can start to combat racism. In this chapter I will explain the origins of race and ethnicity as well as the implications of racism in medicine and health.

The origins of critical race theory (CRT) date back to the mid 1980s which first originated in United States (US) law schools as a re-working of critical legal theory and brought together issues of power, race, and racism to in a contemporary color blind and supposed post-racial society. CRT is informed by, "civil rights scholarship and feminist thought and focuses on dismantling systems that cause racial subordination and injustice," that often lead to disparities in health and socioeconomic outcomes.[1] Only until recently have we started to interrogate the relationship between health and CRT and is evidence by the growing movement of challenging medical algorithms that use race without any scientific evidence. For example, the use of race in glomerular filtration rate has become a topic that has gained a lot of scrutiny and as a result several healthcare systems are starting to abolish the use of race. The reason why it becomes problematic is two-fold, the first is that it either underestimates or overestimates certain risk factors that will determine the care they get and second, it doesn't have any scientific merit as to why and how we use race in these algorithms. The use of CRT in medicine is also starting to become a useful framework to analyzing racial health inequities.

Much of health inequities are marked by racial health disparities. Health disparity commonly refers to a higher burden of illness, injury, disability, or mortality experiences by one group relative to another. In other words, if a health outcome is seen to a greater or lesser extent between two different populations, there is a disparity. These disparities can exist on the basis of race, ethnicity, sexual identify, age, etc. The difference between race and ethnicity is that race is defined as a category of humankind that shares certain distinctive physical traits while ethnicity is more broadly defined as large groups of people according to common racial, ethnic, tribal, religious, linguistic, or cultural origin and background.[2]

To understand how and why racial health disparities happen, we first need to learn how the topic of race is conceived and perceived in America and its relationship to health outcomes with critical analysis. The human race is a mixed product of evolution and biological programming intertwined with the socioecological environment in which we work, live, and grow. Although we do not fully understand the exact all of the mechanisms in which disease manifests itself in humans, we do recognize that our environment is connected to our health. The differences in environmental exposures contribute to poor health and the disparities we see today.

There is often the classic discussion about what is more important—nature or nurture? The truth is that nature and nurture are both equally important as well as critical and central to what defines the intricacies of human health. One can and should argue that because humans share an exact genetic script, and nurture is a prime determinant of survival, the conditions people are exposed to must have the most significant impact on health outcomes undoubtedly. Adverse environmental exposures are prominent in lower socioeconomic areas, which unsurprisingly are mainly occupied by racial minorities. Race has become an arbitrary concept that is displayed as genetic differences and sometimes used against groups to devalue their existence and humanity.

Race is a social construct demonstrated to consistently show links to various consequences and effects in many aspects of our lives. A social construct is an idea that has been created and accepted by the people in a society that serves different purposes. It is an idea that exists not in objective reality, but as a result of human interactions where the humans involved in these interactions agree that it exists. The word race is colloquially used to refer to a person's skin color, religion, or area of origin. Technically, race is centered around people being part of national, social, or cultural heritage. Race operates within two social mechanisms: the first is how you self-identify; however, a second larger factor is how others think you racially identify and how they perceive you. The concept of race has played a significant role in how our society has evolved throughout history. Even today, it influences the way we perceive others and how we experience our lives, most notably, in the form of racial inequity.

Racial inequities in numerous institutions are framed mainly as the result of innate differences and individual behaviors; however, we know that these inequities are mostly reflections of unchallenged social and systemic ideologies. Race is not biological but rather a social construct created by the status quo historically used to marginalize people of color through abusive power. By using modern, rigorous scientific research, experts have scrutinized evidence to come to the consensus that racism causes inequities in our society, not race.[3-7] Although there are inherent differences in our physical appearance due to gene expression differences, they are not causes or reasons for division. This idea is not to say that the solution is to be colorblind.

Instead, we should embrace our differences for the qualities we provide. We don't have to be colorblind to move forward; we can still and should see our color, and we should embrace the strengths in our diverse expressions as a way to overcome racism. Being colorblind in our society is dangerous to our existence as humans and puts humanity's progress at peril.

The bottom line is that we should celebrate our different and diverse cultures; however, only when these differences are used as reasons by the prevailing power structure to manufacture racist policies should we be uncompromising in calling out the problem. Race and the consequences of racism should be acknowledged, addressed, and dealt with in all of their biopsychosocial and ecological manifestations. The heart of the issue is that we have been using race to marginalize and oppress people rather than to celebrate and embrace it. To clarify and reiterate, race is not pathologic; instead, it's the effects of how we use race in our society that can cause health inequities. Moving forward, we must continue to deconstruct the meaning and associations we have with race and reframe how we use race in our society.

Eugenics and Social Darwinism

★ ★ ★

The idea of social Darwinism was central to the formation and justification of the eugenics movement, which started to gain some steam in the early 1900s in the US. The conventional "American" narrative of social Darwinism posits that some people have less fitness than others, and therefore are less deserving of survival. Social Darwinism is ethically, morally, and scientifically flawed and dangerous. The biological use of race—which we have discussed is a social construct—in eugenics is used to decide who is fit to survive and thrive and is just another form of racism in the scientific field. You and your creator can only define your right for survival.

Eugenics is similar to colonialism in the sense that they both aim to subordinate a group of people judged to be inferior. Colonialism refers to the policy or practice of acquiring full or partial control over another country through violent removal of people, occupying it with settlers, and traditionally exploiting the land and people economically.[8] Some examples of European colonization in history include England's colonization over India and the establishment of the British Raj between 1858-1974 or Spain's colonization of middle America and the establishment of New Spain in 1521 which today is known as Mexico. Eugenics on the other hand is more of a social and political philosophy that is focused on improving the genetic quality of a human population, historically by excluding people and groups judged to be inferior or promoting those judged to be superior.[8] Both of these concepts are connected together by one underlying mechanism which is racism. Eugenics started to become popularized because it provided a "scientific" mechanism to justify racism and racial superiority.

Scientific racism is the belief that empirical evidence can support and justify racial discrimination and racial superiority or inferiority - one such proponent was German physician Dr. Johann Friedrich Blumenbach.[9] In 1779, Dr. Blumenbach was the first to divide the human species into five races based on his descriptions of sixty human crania, which led to the study of anatomical variation in different human races. However, in all of his research, it was never concluded that racial differences within the human species led to any biological inferiority based on the anatomical variation.[10] Even during my medical school education, we studied every detail of the human anatomy of twenty-three cadavers to learn about anatomical variation. There are cases where anatomical variation can threaten survival, but most time, these variations are benign. Biological inferiority in the human race is a dogma with roots in racism, colonization, and eugenics.

Still, eugenics in America was popularized in the mid-1800s and was practiced until 1944, led by doctors who practiced scientific racism. Many argue that it is still happening today. In the early 1900s, eugenics spread to Europe, mostly informed by white scientists in the US. In America, we encouraged it through forced sterilization, segregation, and racist anti-mixing laws, all with the objective of the extermination of non-white people.[11] In Germany, Nazi doctors took inspiration from American scientists and practiced eugenics by slaughtering, torturing, conducting experiments, and murdering millions of Jewish people based solely on these differences.[12] All falsely conducted under the veil of science, these crimes committed against humanity tended to be orchestrated by distinguished scientists inspired and actively influenced by eugenicists in America.

It is vital to understand the historical context of eugenics for several reasons. The first is that we must realize we are worthy of becoming a distinguished scientist, no matter what background we represent. This is important because it offers a way to counter and challenge these negative and harming ideas that often get popularized in racial groups with significant homogeneity. The second is that because there was a lack of diversity in science during that time, these dangerous ideas were popularized by white scientists and went unopposed. Because the practice of eugenics was unchallenged, it became common practice and further marginalized minorities as time went on, discouraging diversity and inclusion in the field of medicine. It is crucial to understand that much of the past research to inform the ideologies we have in our modern society was inspired by the ideas of patriarchal, cis-gender, white, male colonizers who were power-hungry and greedy. This assertion is not a sweeping indictment to say that all white people are evil and racist because we know that is not true. However, we understand that these were the people who created and designed the ideas that still affect us today.

An FGA will realize that just because your family comes from whatever nation outside of the US, or just because your skin color is not "white," it does not mean you are inferior. Immigrant, Black, and Brown are just as beautiful and equally worthy of respect.

Saying that people from the African diaspora are inferior to those from a different geographical area based solely on their skin color is the same as saying people with blue eyes are inferior to people with brown eyes. On the contrary, having darker skin actually protects you from harmful solar rays from the sun. There is no empirical evidence highlighting how differences in diverse phenotypic social characteristics are deleterious or to justify white superiority.[13,14]

White superiority or supremacy refers to the idea that the White race is believed to be superior and therefore should dominate every aspect of society, typically to the exclusion or detriment of other racial and ethnic groups.[14] The concepts highlighted earlier such as eugenics and colonialism operate under these ideas and beliefs. Currently, the status quo of race in America is seen in the superiority of one race over all others. America's flawed interpretation of race—used as the basis for systemic racial inequities—is a status quo that is not accurate and that we must challenge and dismantle.

Today, social institutions such as the medical, immigration, criminal justice, and education systems all still retain some elements of eugenics. For example, cutting social support programs including health insurance and benefits that disproportionately impact the poor; tearing apart families and putting kids in cages, the low-income and English as a second language kid being pushed out of the educational system, and the racial profiling and mass incarceration of Black and Brown people. As a country, we must loudly condemn the misuse of racial differences as a justification for such marginalizing policies. Our differences do not determine any innate or biological inferiority, and thus we should not suffer different consequences from our counterparts based solely on the factor of race. There is no "master race." There is the human race and the several racial, social categories under it in equal measure.

The New American thought must understand that the dominant power structures under which we operate cannot have the power to tell us who we are and where we stand. This thought means that we will no longer allow others to assign meaning to us based on differences that they fail to understand. The differences we find in our diversity are an advantage we can use to advance our world to brighter and more equitable horizons, and not a means to oppress another group based on wicked science without merit. Instead, we should assign value to our species by using cultural responsiveness with fundamental human rights principles and prevailing evidence-based, multidisciplinary guidelines that have undergone extensive, rigorous research and approval by communities. Only then can society move toward justice, equity, and prosperity.

Implicit Bias and Mental Heuristics

★ ★ ★

Human behavior is complicated, so we all use mental shortcuts, also known as heuristics, as a form of decision-making in our daily lives. These cognitive heuristics are happening at all times and are often obscured by bias. This phenomenon is innately human. However, it is imperative to understand that heuristics can also be dangerous and inaccurate, primarily when conscious and unconscious biases influence your decision-making. As humans, we all have preferences, including a natural bias toward specific communities, languages, locations, and food. While having preferences and biases is not always ill-intended, the consequence of expressing certain tendencies can differ vastly from the original intention and either overtly or covertly cause damage to others. We can think about implicit and explicit bias not as a negative social mechanism but as something that continuously challenges and encourages us to grow. It is more a matter of how you manifest these biases and preferences because they can be productive, effective, and in some cases, lifesaving.

For example: Let's say there is a thirty-seven-year-old Black female who is postoperative day two from a transverse C-section with a history of gestational diabetes that was complicated by postpartum hemorrhage. She is now experiencing maternal fever and difficulty breathing. What do you do next? Based on the statement above, as a physician, I must be able to take mental shortcuts backed by evidence to ensure that I can provide compassionate and culturally responsive health care backed by science. Consequently, these shortcuts can save a life. They also allow me to consider her race as a possible risk factor for inequitable treatment in a system built and designed by racists, evidence by the documentation that Black mothers are more likely to die from postpartum complications.[15]

In medical school, we are taught that race is a singular risk factor for several conditions and inequitable health outcomes. This idea is an easy mental shortcut for diagnosis. Still, it is unhelpful insofar as it reinforces associations without ascertaining the root cause of diseases that are caused by social factors. For example, suppose we are talking about maternal mortality, triple-negative breast cancer, or sarcoidosis. In that case, we are taught that being an African American or Black female puts you at a higher risk of having those conditions. The lesson ends there without any other explanation of why we see these associations. Here's another example: if we hear about diabetes or obesity, we automatically associate it with Hispanic/Latinx people. But why are Latinx people experiencing higher rates of obesity than their white counterparts? Is it because of biological reasons—that they are merely Hispanic/Latinx? No, we have already seen how such beliefs are rooted in racism with no empirical evidence. Correlation is not causation. Instead, it is probably because of socioeconomic and environmental factors—perhaps the neighborhoods they live in are unsafe and do not have a sidewalk that allows them to get to a grocery store. Maybe there is no neighborhood park where they can exercise; perhaps they do not have the economic means to access healthy foods.

Although there has been evidence to show a genetic predisposition to certain conditions, we know that environment is a stronger predictor. For several years during school, healthcare professionals have only been taught to reinforce associations and not trained about or equipped with the tools to target the root cause of why race is considered a risk factor for different diseases. And as a result of failing to learn about these root causes and how we can address them, health disparities will continue to exist. Is there a better explanation for why a race, a social construct, puts you at higher risk for certain diseases? There isn't a clear answer. However, we know that it is not innate; instead, it reflects the interaction between the patient and their social and physical environment.

Racism is Pathologic; Race is Not

★ ★ ★

Contemporary science and our history have taught us that the perversion of racism is harmful, dangerous, and violent against the powerless and people of color. It is essential to understand the biopsychosocial origins of race in America because of how race has been used in medicine and portrayed throughout its inception. Biopsychosocial mechanisms refer to the biological, psychological, and social impacts that the social construct of race has on humans. All three are interconnected and seen as a whole instead of as its disparate parts because we know that health is a product of all of these areas. Moreover, we know that health is not just the mere absence of disease or pathology; it also includes well-being in all aspects, including physical, mental, spiritual, and emotional well-being. One can't merely use just one of those aspects when we are talking about the impacts of racism because they all coexist together and influence each other. Your physical health is connected to your mental health, which is related to your emotional health.

Race has been weaponized in the US and has not always only impacted minorities. On top of the African slave trade, the British also oppressed the Scottish and the Irish and even enslaved them as indentured servants in Europe, however, when they came to the US, they became "white" over time.[8] In other words, America assimilated many Europeans to the belief of white superiority. In the US, race was developed as a social category to create a caste system that put minorities on society's margins. As a social category, it has been used to justify several human rights atrocities throughout history. It is no longer okay not to be racist; now, we must be anti-racist.

Structural pathologies are the differences in societal systems that generate health disparities by making individuals more susceptible to certain diseases. One example of a structural pathology is how race continues to be an indicator and proxy of socioeconomic health outcomes in the current American health care system with the transportation system. These structural pathologies are directly harming people when highways and high-traffic areas are built in low-income communities where people of color predominantly live and where residents are more exposed to environmental pollutants.[16] By allowing this type of development in such neighborhoods, we are directly impacting the health of the people who live in that area. For example, kids who have a chronic disease like asthma are more likely to go to the emergency room when environmental pollutants further exacerbate their asthma. Considering that they belong to a low-income community, it is also probable that they might be uninsured, which means their care ends up being paid for by our tax dollars. Indirectly, these highways and high-traffic areas can also have more global consequences as they contribute to climate change, such as the increased frequency of climate disasters, leading to adverse health outcomes for the broader community. Racism has direct and indirect effects on us in several other aspects. It doesn't just affect us based on where we live and what we are exposed to in the environment.

All this being said, we must acknowledge that race itself is not pathologic; rather, how we treat people based on their race—racism—is. I am not arguing that the concept of race is terrible or that it doesn't exist because, as a social category, it has been a helpful way to self-identify, and it is also ingrained in our everyday lives. Instead, I am acknowledging that racism—the unequal treatment of others based on race—is pervasive in American society, as it has been from the beginning. It, directly and indirectly, causes unequal health outcomes.

It is important to remember that white people cannot experience racism in a system that was built and designed by whites. I am not saying that prejudice and bias don't exist against white people. To clarify, I do believe white people can experience discrimination in the form of bias or prejudice. However, this is not the definition of racism. Racism is defined as the systemic oppression of other racial groups by the dominant power structure, in our case, white supremacy. We don't live in a society where every racial group has an equal power status or similar opportunities, even though people like to believe this is the basis of America. In the particular context of American history, white people have historically benefited, even centuries later, from the enslavement, murder, and dehumanization of people of color. We live in a system designed by white people. By virtue of this system, white people continue to benefit from others' oppression despite individual behaviors or intentions. Because of this, white people cannot experience racism. In other words, reverse racism does not exist.

I believe that interracial prejudice and bias exist, mostly outside of the majority power structure, precisely prejudice, and bias among different minority races. And I want to acknowledge that it can happen at various levels, including at the internal, community, and systemic levels. That being said, you cannot be racist in a society when you are outside of the dominant social group that has historically abused power at the expense of the people in your community. Some people argue that white people will soon become the "minority" in America to counter this argument. I think this is false because being a minority does not only mean being a smaller subset of the population; it also means not having equal representation in positions of power, such as in our government. Being a minority is not about numbers; it is about the lack of power.

At the expense of others, institutional racism grants privileges to the dominant racial group. These privileges continue to be inherited by their future generations, leaving other communities in the margins. To understand racism, we must examine the origins of race in the context of historical and geopolitical events throughout human experience. If we do not begin to repair the harm we have caused throughout history, we will easily fall into the belief that what happened before is not impacting us now. Without adopting an approach of CRT, we will be unable to have an informed discussion of the implications of race and what to do next.

Color Blindness Is a Form of Modern-Day Racism

★ ★ ★

I still remember the day I was walking home from college when I was a freshman. A police car was driving past me, and as soon as we crossed paths, he immediately busted a U-turn in the middle of the street. The officer pulled up behind me and asked me to stop. I was frightened just because I wasn't sure why I was being pulled over. As he came closer, I asked him what I had done wrong, and he said that I looked suspicious. But he wouldn't say what I was doing that was so suspicious. I had already experienced racial profiling when I was younger, but I was disappointed that this was still happening to me, even in college. I couldn't help but think that he had probably pulled me over because I was a Brown kid. Infuriated, I yelled, "Is it because I am brown?" The officer quickly scoffed and said, "Kid, I don't see color. I am just doing my job."

Colorblindness is a concept that promotes the idea of being blind to our racial and ethnic differences. However, this practice does not lead to racial harmony if we ignore, overlook, and dismiss racial and ethnic disparities. Instead, it denies the very existence of the diversity that people have to offer. The increasing diverse racial demographics in our country are becoming more and more critical. Instead of saying we don't see color and pretending that race does not exist, we should reframe our understanding to acknowledge our differences and the rich diversity among our racial and ethnic groups because they do matter. Learning about others' cultural experiences, even internationally, will allow us to move forward effectively and collaboratively as a country. For example, learning about how other healthcare systems operate based on their cultural norms is essential for the US because we are diverse. Therefore, to meet the needs of our growing diverse population, we must be able to see what works for different people instead of assuming that everyone wears the same size shoe.

Additionally, instead of adopting colorblindness or even going to the opposite end of the spectrum to promote cultural competency, we must strive to be culturally responsive, as I mentioned in chapter one. It is nearly impossible to be culturally up to date in any culture, considering how fluid culture is and how it keeps changing over time. Having such new and varied perspectives will help us better understand the strengths that we all bring to the table. The idea that we do not see color is problematic for several reasons. Not only is colorblindness scientifically, morally, and ethically flawed, it can also damage our diversity. Being colorblind means being ignorant of people's cultural differences. We miss the opportunity to learn how unconscious biases can influence our decisions, expectations, and behaviors when working with different people.

Colorblindness is modern-day racism because of the implications it has on people of color. Historically, we have seen various forms of discrimination, and it has recently evolved into what we see today in our current social context. Diversity can be damaged by colorblindness because we don't see color; we are not acknowledging our differences and unique narratives. The fact remains that we are not a colorless society. We all have a rich genetic diversity that is beneficial to our survival as human species. Being colorblind is no longer an option. If you are colorblind, you are just lying to yourself and saying that you cannot acknowledge my Brown skin and the historical trauma my ancestors and I have endured.

Until this day, my encounter with that police officer sets off several unanswered questions for me. Mostly, why is our policing system set up this way? I know those police officers have the right to defend themselves when they feel threatened, but my skin color should not be something that makes them feel threatened. The officer's comment didn't sit well with me, mostly because I know he was either probably lying about not seeing color. Because the truth is, we all see color, especially after considering how ingrained it has become in American society since the founding of the country. I would soon find out that our modern-day police system was just an evolution of the slave patrol created after abolishing slavery. To begin understanding who we are as humans and how and why our society operates the way it does, we must continually ask ourselves questions. The goal is to have a critical appraisal of what we believe and how we behave.

In summary, I discussed how race works as a social construct and the implications of living in a racist society have on people. In this chapter, we discussed how the dominant power structure has abused science and how the concept of race has been used historically. Its systemic manifestations still go largely unchallenged today. In the next chapter, we will learn more about how race is only one part of the equation determining health outcomes. Chapter three will discuss the basic idea of the social determinants of health as they pertain to identifying the root causes of structural health inequities in health care and building a society that will someday be able to rid itself of racial health inequities.

REFERENCES

1. Martinez AY. Critical Race Theory: Its origins, History and Importance to the Discourses and Rhetoric of Race. Frame NO. 27.2 November 2014.

2. Watt S, Norton D. Culture, ethnicity, race: what's the difference? *Paediatric nursing.* 2004;16:37-42.

3. Braveman P, Gottlieb L. The social determinants of health: it's time to consider the causes of the causes. Public Health Rep. 2014;129 Suppl 2(Suppl 2):19-31.

4. Bailey ZD, Krieger N, Agénor M, Graves J, Linos N, Bassett MT. Structural racism and health inequities in the USA: evidence and interventions. Lancet (London, England). 2017;389(10077):1453-1463.

5. Williams DR, Lawrence JA, Davis BA. Racism and Health: Evidence and Needed Research. Annual review of public health. 2019;40:105-125.

6. Paradies Y, Ben J, Denson N, et al. Racism as a Determinant of Health: A Systematic Review and Meta-Analysis. PLoS One. 2015;10(9):e0138511.

7. Cobbinah SS, Lewis J. Racism & Health: A public health perspective on racial discrimination. J Eval Clin Pract. 2018;24(5):995-998.

8. Okrent D. The guarded gate: bigotry, eugenics, and the law that kept two generations of Jews, Italians, and other European immigrants out of America. First Scribner hardcover ed. New York: Scribner; 2019.

9. Paul DB. Darwin, social Darwinism and eugenics. In: Cambridge University Press; 2003:214-239.

10. Rupke NA, Lauer G. Johann Friedrich Blumenbach: race and natural history, 1750-1850. Abingdon, Oxon;New York, NY;: Routledge; 2019.

11. Black E. War against the weak: eugenics and America's campaign to create a master race. New York: Four Walls Eight Windows; 2003.

12. Fischer BA. Maltreatment of People With Serious Mental Illness in the Early 20th Century: A Focus on Nazi Germany and Eugenics in America. The journal of nervous and mental disease. 2012;200(12):1096-1100.

13. Sensoy Ö, DiAngelo RJ, Banks JA, Columbia University. Teachers C, Teachers C. Is everyone really equal?: an introduction to key concepts in social justice education. Second ed. New York: Teachers College Press; 2017.

14. McGettigan T, Smith E. A formula for eradicating racism: debunking white supremacy. New York: Palgrave Macmillan; 2016.

15. Shahin Z, Hardwick I, Jeffery N, Jordan J, Mase W. Maternal Mortality among African American Women in the State of Georgia, Causes, Policy, and Ethical Considerations. Journal of the Georgia Public Health Association. 2020;8(1).

16. Zimring CA. Clean and white: a history of environmental racism in the United States. New York: New York University Press; 2015.

The Social Determinants of Health

A basic foundation to understanding public health sciences

"Achieving health equity requires valuing all individuals and populations equally, recognizing and rectifying historical injustices, and providing resources according to need."

—*Camara Jones*

My Experiences with the Social Determinants of Health

★ ★ ★

I grew up as the son of two formerly undocumented residents of the United States (US), which has exposed me to many social determinants of health that have been adversities and humbling experiences. My parents came to the US to search for opportunities and as a means to escape from more drastic and deadly disadvantages at home. No running water, electricity for only a couple of hours at night, and crowded small rooms were an everyday reality for my family in Mexico. Due to our limited number of resources, no one in my family before me graduated from high school and went to college. In fact, the furthest that anyone had made it was a sixth-grade level of education. The expectation was that you had to work to put food on the table before doing anything else. This way of living was their means of survival. This is a reality for my family, and it still is for many people worldwide.

I was born in East Los Angeles (LA), California, and spent some time in an underserved urban community. As a child, I was regularly exposed to poverty, violence, drugs and many adverse childhood experiences. Gangs terrorized my family, but my family was reluctant to report it to the police as they were afraid of being harassed by the police, or worse—getting reported for deportation. Buying groceries sounds like a routine activity, but it meant the risk of being separated from my family. Authority figures, including police officers, would discriminate against my family and threaten to deport them because they "weren't real Americans." We were in limbo, with nowhere to turn but to escape once again.

It didn't take long for my family to move to a rural town at the top of the Rocky Mountains. Since I was so young, LA's transition to a rural village was not difficult because I didn't really have friends. I attended an underfunded school and medically underserved town with some of the highest dropout and teen pregnancy rates in the state. I was surprised I even graduated from high school because I had been told many times that I would not graduate. At first I believed what people would tell me, however I was able to push through and graduate high school. Even as I was graduating, the cost of getting an education was a scary thought. I was fortunate enough to earn scholarships to attend college. Otherwise, I probably wouldn't have had the opportunity to pursue an education.

Our financial circumstances were so meager that my family lived in a trailer park about 3 miles away from the town. As a consequence of our low-income status, we were forced to live farther away from the school, grocery store, and health care facilities without any access to transportation. These circumstances significantly contributed to my grandmother having uncontrolled diabetes and hypertension. Unfortunately, my grandmother was not the only victim of these inequities—I have seen many community members buried early in their graves for similar reasons.

All of these different determinants have had catastrophic consequences for my academics, finances, and mental health. Despite the odds, I persevered in the face of adversity. I am not offering this narrative to elicit sympathy for my journey, but rather to highlight some of the determinants of health I had to face in my life. I want to provide insight into my character as someone who has survived these overwhelming circumstances, learned from them and translated them into personal growth for the greater good of humanity. Through these different experiences, I learned different ways that health is more than just a doctor's visit.

My experiences have shown me that calling for health equity globally should be central to the mission and vision of every organization that believes in the social determinants of health. My goal for this chapter is to provide a basic framework for health and to inspire diverse perspectives for collaboration among medical, legal, educational, and scientific professionals to better understand the impact we can have using the social determinants of health.

What Are the Social Determinants of Health (SDOH)?

★ ★ ★

The concept of social determinants of health (SDOH) is a tool used in public health science to study the causes and factors that determine health outcomes. These health outcomes can range from things like heart attacks, strokes, hospital admissions, and even death. It has been an emerging topic in medicine to help better health professionals identify risk factors and treatments or prevention strategies. In its most basic form, the SDOH are factors that contribute to your health status outside of the healthcare system. The factors include race/ethnicity, job status, education status, access to transportation, access to healthy foods, and policies/laws set by people in power. All of these factors have been just as, if not more, important than the other factors we tend to associate with health outcomes linked only to individual choices. The SDOH goes beyond just individual and behavioral conditions that cause disease and health outcomes and highlight how individuals interact with environmental, sociopolitical, cultural, and other factors that contribute to health outcomes.[1]

When we talk about the SDOH, we will need to discuss fundamental causes and proximate causes related to the disease. Etiology is just another term for "direct cause." Let's take strokes, for example, and understand it in terms of fundamental and proximate etiologies of disease. Fundamental causes of strokes include not having the money to buy healthy food, not having insurance to purchase medications, etc. On the other hand, a proximate cause of strokes would be factors that directly cause the disease, such as having a genetic predisposition, elevated blood pressure, coronary artery disease, etc. Knowing the fundamental and proximate causes of disease is essential because they allow us to study health outcomes as a sequence of events rather than just a single and simplified outcome.[2] These become important when we start talking about casual pathways, which help us better understand what causes disease.

There are different ways in which risk factors can affect disease outcomes. Proximate risk factors are sometimes framed as behaviors usually within the individual's control, such as smoking and eating habits and wearing helmets. This connection implies that people are to blame for their health, which is partially true. However, we must also understand that these behaviors unfold within a bigger context. Proximate risk factors must be contextualized to realize both their individual impact and population impact. We also now understand that there are fundamental causes that are also responsible for adverse health outcomes. Different fundamental causes that can cause harmful health outcomes include stress, genetics, culture, and social influences from preexisting conditions in the environment.[2] This risk exposure can affect people's health and the onset of disease through biological mechanisms and lack of access to necessary financial or social resources to maintain health.

One easy way to better understand the SDOH and different fundamental causes or proximate causes is by using an analogy of a rapidly moving river stream, like the ones you can see after the snow has begun to melt from the mountains. Now, imagine that you are standing at the bottom of the mountain where the river starts to split, and you see someone drowning in that river. In this case, after assuming it is safe to do so, the best thing we can do next is rescuing the person from the river. Now, let's say a couple of minutes later, you see two, or three, people in the river. Luckily, you have extra hands so that you can rescue them. However, now you see another person, and you know something is up. A public health professional will ask: wait—why are people falling in the river in the first place? It turns out that there is a broken bridge just upstream of the river, and people keep falling because the bridge has not been fixed. So, the next step would be to improve the bridge so that people don't fall into the river and so you can prevent any of these incidents from happening in the first place. Now, in this case, we know that falling into the river is the wrong outcome. The proximate cause is walking over the bridge and falling in. When we think about fundamental causes, we then look at why the bridge is broken, to begin with. Why is it not fixed yet? Should we close down the bridge because it is not safe? Using this analogy, you can have a clear example of fundamental causes versus proximate causes of disease.

Casual Pathways and Health Equity with the SDOH

★ ★ ★

Fundamental and proximate causes of disease are essential to explaining how casual pathways occur and how they could help us address a patient's harmful living conditions. Causal pathways and disease etiology allow us to understand some disease patterns and better approximate risk factors that contribute to disease. By understanding disease etiology in the SDOH, we can understand and identify the causes or origins of disease by examining all factors contributing to health outcomes, including social ideologies rooted in history. An ideology is similar to construct in the sense that it is a set of shared beliefs within a group, which influence the way individuals think, act, and view the world. It is crucial to understand which social ideologies are either causing or preventing health inequities because they allow us to trace the impact social factors have on disease development throughout the patient's life course. Using this framework, we can understand how the SDOH provides a comprehensive approach to understanding health equity.

There are possible causal pathways that show us how social structures can affect health status. A causal pathway is a precise mechanism in which an exposure leads to a disease, where the development of disease has multiple causal pathways. Some of the causal pathways we see where social structures can determine health status include the physical environment, the social environment, individual lifestyle, and differential access to healthcare services. When we see the connection between the various causes and risk factors, and causal pathways of a disease, we begin to understand how the SDOH describes many health inequities in society today. Some of the leading causes of death in the US include heart disease, cancer, chronic lower respiratory disease, strokes, Alzheimer's disease, diabetes, influenza, and pneumonia.[3] We have seen that cardiovascular death rates per 100,000 by race and sex have declined over time; however, it has not decreased at the same rate for everyone. In particular, as we see better outcomes in the general population, we also see that there are still large disparities between racial groups.

The SDOH are directly influenced by public policy decisions made by governing officials. Most health policy focuses on healthcare treatment after diseases have occurred, but this area only accounts for 30 percent of outcomes.[4] Using the SDOH framework allows us to focus on the rest of the 70 percent contributing to outcomes.[4] More important, it helps us examine the root causes of many social inequalities. Preventing the emergence of a disease requires using health policy to address the unhealthy social conditions we already live in.

To obtain equity in health, we must look toward health policies that are mixed with social policy. We can begin reshaping the SDOH by viewing policy through a health equity lens. This means that health impacts should always be considered when making decisions that have social and political effects. Furthermore, these impacts should be carefully studied, and risk/benefits must be known, epically among minority groups or anyone else who are directly impacted. Every aspect of government and the economy can affect health equity, including finance, housing, transport, and health, to name a few.

Populations that experience more chronic stressful situations receive multiple adverse health effects that lead to a state where your body's natural stress hormones are always elevated.[5] These quick and long stressors can originate from different direct and indirect factors that lead to constant activation of your nervous system responsible for automatic functions. Your body is always on full alert automatically, resulting in the allostatic load. The allostatic load is a condition where your chronic stress system is continuously activated.[6] It causes health outcomes such as blood pressure changes, higher mortality rates, cardiovascular disease, stroke, and the loss of physical and cognitive functioning.

One prime example of the stressors that lead to the allostatic load is socioeconomic status, such as neighborhood poverty in low-income, multiracial, urban communities. Neighborhood poverty is related to wear and tear on organ systems and mental stress, showing that chronic stressful situations directly affect the human body. For example, the lack of cheap and accessible transportation in a neighborhood separates low-income people from healthcare facilities. This forces families to spend a large number of their budgets on cars, leaving them struggling to save money for other expenses. Although they can travel because they have a car, the car's purchase is at the cost of different needs, such as buying healthier food and obtaining health care. If they do not have a car, they must rely on walking if there is no public transportation. However, some of these places have un-walkable areas while also lacking bicycle-friendly roads. In this situation, financial stressors contribute to the adverse health outcomes that lack cheap and accessible transportation.

How to Study the SDOH

★ ★ ★

The social determinants of health engage several factors that are beyond those of only biomedical or behavioral risk. Many of these determinants include early life, education, employment and working conditions, food security, gender, healthcare services, housing, income distribution, social safety net, social exclusion, unemployment and employment security, social support, and race. There are different social, economic, and political resources and structures that influence health and several of the conditions in which people are born, grow, live, and work. All are interrelated with the distribution of money, power, and resources at global, national, and local levels, which give rise to health inequities and unfair and avoidable differences in health status in countries and populations across the world. Every one of us has lived in a world where the social determinants of health have impacted our own lives and those of others, including family, friends, and the community.

Without a coherent approach or guiding conceptual framework, we're just chasing after risk factors in unexpected ways with no clear outcomes to examine. We want to study the social determinants of health beyond collecting experiences and use theory and conceptual frameworks to understand how we can begin to understand the relationships between the SDOH and health outcomes. Having this framework gives us the power to think critically and form critical research questions. The importance of this theory is to explain or predict the patterns that we observe. Still, we need explanations for why and how we think they occur, which is a practical application of theory.

In the next section, we will be examining how the SDOH can be studied in different settings. I will be using various case studies to demonstrate how the SDOH is manifested in our lives. In part two of the book, we will also translate these theories into practice by looking at the life course. The life course highlights cumulative risk, biological susceptibility, and how our social context impacts our health outcomes, and it is another tool that can help us study the SDOH. It is also a useful tool we can use to illustrate the fundamental causes of disease that are structural pathologies and the inequitable distribution of resources. The best setting in which to study the SDOH is your backyard and your community.

Neighborhood Analysis and the Impact of Redlining on Health

★ ★ ★

One way you can examine the social determinants of health is by doing a neighborhood comparison analysis. The data collection can include canvassing with a community checklist or using Google Street View to conduct a built and physical environmental scan. Examine the different types of resources available to the community, including access to necessary resources for life or maintaining a healthy lifestyle, such as having access to an area for physical activity and having access to healthy food. These different resources can include grocery stores, gyms, sanitary services, schools, and safety emergency services. Using this approach provides you with a systematic way of analyzing the neighborhood.

The Google Street View tool can give you a snapshot or an idea of the neighborhood's physical environment. You can also observe the broader community since you can zoom in and out to different parts of the area instead of walking for hours. However, the tool does not provide much direction in terms of specific highlights or features of the site that you can look at since Street View consists only of snapshots in time that may not directly represent the neighborhood's lived-in actions. Additionally, Google Street View may be outdated and may not describe what is going on at the current moment. The benefits of doing community canvassing are that you can see the built environment more accurately. However, doing community canvassing is not always an option depending on where you need to go and sometimes does not allow time to explore the entire neighborhood you are investigating. Using both tools in conjunction can be useful and efficient to help you examine the social determinants of health related to the physical and built environment in your community.

One particular element you can focus on is food security in your community, including the accessibility to gas stations and grocery stores. Food availability plays a critical role in the neighborhood's environment and health outcomes. Food is not equitably accessible and available in neighborhoods according to household income level and race, contributing significantly to poor food environments within marginalized neighborhoods. In turn, food stores offering healthy foods are generally located low income areas and are concentrated in neighborhoods with higher household incomes. This social mechanism of neighborhood segregation in the US has led to many structural conditions that result in reduced or low health due to the lack of nutrition in a community based on its demographics.

An example of how socioeconomic and racial neighborhood segregation came to be in the US is through redlining. Redlining was a common practice in the US and began in 1934 when the National Housing Act was established and created the Federal Housing Administration. Redlining was a practice where the systematic denial of various services to residents was based on neighborhood demographics.[7,8] One example of redlining is in racist moneylending practices where banks refuse a loan or mortgage to homeowners or business owners of a specific demographic from a particular community. This gives rise to segregated neighborhoods that continue to be segregated over generations. Redlining keeps racial minorities within their poor neighborhoods that have historically been underserved while keeping certain areas reserved for white people. So, why is there a lack of access to proper food nutrition in more impoverished neighborhoods? One reason is that these communities have been consistently denied amenities and resources that would have enabled their success. The systematic denial of essential resources and services by federal, state, and local agencies, including the private sector, causes displacement and the exclusion of people from equity, putting these already underserved communities at a higher risk of poor health outcomes.

Hopefully, it is now evident by this point that where you live can impact your health. We also know that racial minorities have been strategically placed in neighborhoods with undesirable living conditions because of redlining. Several studies have uncovered and shed light on these systemic inequalities in neighborhood development and health.[9-11] For example, a recent analysis of an important study found that one of the most important influences on health is your neighborhood socioeconomic status.[12] Meaning that just by knowing your zip code, we can predict how long you are going to live and what kind of health conditions are likely to be common in that area. Research also shows that adolescent development in neighborhoods with higher socioeconomic status shows higher positive health outcomes in behavioral and emotional health, less risky sexual activity, and less teenage pregnancy. This is likely because these areas have more access to educational services and after-school programs.

Conversely, children living in more impoverished areas have poorer health and education outcomes, resulting in risky sexual activity and higher teenage pregnancy rates.[13] Neighborhood environments with lower socioeconomic status may have a higher proportion of the younger population, leading to higher teen pregnancy rates and other adverse socioeconomic health outcomes. When people have poor health outcomes, they are also more likely to lose their financial stability, resulting in a vicious cycle of generational marginalization, which is part of the systemic inequity present in such neighborhoods.

As a result, this environment sets up a systemic cycle of poor health outcomes for generations. When these teen mothers give birth, they usually have high rates of low-birthweight babies; meanwhile, low-birthweight babies tend to be teen mothers' children.[13] In turn, children who are born as low-birthweight babies have a higher chance of becoming pregnant teen mothers themselves because of the social conditions in which they already live. After years pass, their children have a higher chance of teen pregnancy, leading to more low-birthweight babies and continued high teenage pregnancy rates. Even after controlling for every other variable, race is associated with lower birth weight in African American females than their white counterparts.[14,15] This structural pathology exists because of the systemic inequality that has resulted in segregation, displacement, and exclusion from society instead of something that is intrinsically biological.

Environmental Justice and Built Environments

★ ★ ★

Environmental justice is an approach that focuses on protecting people's rights and health from toxic ecological exposures or conditions.

Environmental justice covers several different fields, including health equity, law, politics, and history. A built environment is an area in where people live and it also is an area that takes into account the structural surroundings around the area.

Using the upstream and downstream mentality we learned in an earlier section, let's examine this case study where you have a Native American tribe living in a tiny village with limited access to clean water. Again, let us remember that Native Americans were forcefully removed from their lands and placed in the living conditions that they reside in today. It turns out that the only things they have around their neighborhood to eat is McDonald's, Burger King, and Wendy's. The closest store is thirty minutes away by car, but community members don't have access to transportation. They also have poor sanitary conditions due to the exploitation of a factory built upstream from the river, contributing several pollutants to the water. Yet, due to political reasons, authorities fail to address this public health issue. Native American community members are forced to live in an underserved area downstream from factory pollutants due to environmental injustice.

The built environment facilitates the designing of transportation and has an essential impact on many other areas. In 2011, the Leadership Conference Education Fund assessed that access to public transit is a civil right and examined the vital role that transportation plays in determining health outcomes. They found that a lack of affordable and accessible transportation is a significant contributor to health disparities because, for one, it isolates low-income families from getting transportation to a place that benefits their health, such as a doctor's appointment or a job. Underinvesting in walkable communities, transit buses, and bicycle-friendly roads contributes to many other health issues for these neighborhoods other than just a lack of transportation. Transportation has a significant impact on many different aspects of health that may not be so obvious. The only way to change and improve these neighborhoods' health is to incorporate accessible, affordable public transportation.

Rather than building affordable public transportation in these locations, however, there is a focus on building highways and freeways in the area that serve private vehicle owners. These communities become areas where private transportation is common but is not available for community members. This results in residents living in more cars and increased traffic, especially in the communities built right next to freeways. The more cars there are on the roads, the higher the low air quality levels associated with higher asthma rates.[16] Obesity rates are also becoming an issue for these communities because they do not have access to sidewalks for biking and safe, walkable areas, which provide more avenues for physical activity.

Suppose we are to incorporate environmental justice elements in underserved neighborhoods already with poor health outcomes. In that case, one way is to design these neighborhoods to promote physical activity, ensure food security, and economic opportunity to reduce the risk of diseases like heart disease and obesity. Individuals and families who are at a higher risk for obesity-related diseases live in neighborhoods that don't have walkable access to public parks or wellness facilities compared to other groups. Their communities are also predominantly served by fast-food options. These are just a few examples of how obesity may have more fundamental causes than proximate causes and how built environments can play a crucial role in determining negative health outcomes for the community and further driving more adverse health behaviors. For example, the closer a school is to the community, the more likely people will participate in active transportation, such as walking and biking to and from their destination. However, if the school is located farther away and the neighborhood is less walkable, people will not take the bus. Even if a person who lives far away decides to walk to school, they can only do so if walkable areas like sidewalks are available.

It is critical to design environments with an environmental justice approach because of how the physical environment has been historically built to benefit some groups. For example, a lower-income neighborhood with high obesity rates may not have walkable access to a park or healthy food options available. Meanwhile, right across the town, there is a predominantly white neighborhood that has access to better open spaces and more nutritious food, and that has lower obesity rates. These fundamental causes of obesity are rooted in redlining, which we highlighted earlier in the chapter. All this to say that there isn't a fair distribution of amenities for everyone, especially considering the social context of how the people who live in underserved neighborhoods have historically been marginalized.

Several other factors contribute to the lack of physical activity in the community. There are also reports of higher crime rates in underserved neighborhoods compared to anywhere else.[17] This concept should not be confused only with community violence, which is a smaller subset of crime, but also the constant harassment and executions committed by the police force. So, even if there were better access to walkable areas and neighborhood parks, the lack of safety would limit community members. Though access to walkable communities encourages physical activity, if there is high crime, then people won't want to walk or bike to work. This scenario causes even more barriers to overcome health disparities in low-income neighborhoods regardless of their built environment. As a result, we sometimes still see that having access to amenities does nothing to change the rates of obesity. All this may suggest that the higher prevalence of obesity and obesity-related diseases in lower-income areas result from either not having access to places for physical activity or not feeling safe enough to participate in activities even if they have access.

How We Can Use the SDOH to Eliminate Health Inequities

★ ★ ★

Using the social determinants of health as a framework, we can understand how structural pathologies and racial health inequities occur in our lives. Where you live in the leading indicator of your health, and your health determines your ability to thrive in society. If you are not healthy, you are less likely to have access to education and more likely to have a poor-paying job that will, in turn, not be sufficient to sustain you to live a healthy life. This cycle of poverty and marginalization has to be addressed because it is harming our future generations. We should care about this issue because the SDOH impacts people, whether we are talking about the children of Latino immigrants, Black neighborhoods terrorized by the police, or the Asian and African refugees leaving their homes to come to the US in need of our support. These populations will not get smaller; they will only continue to grow in the US.

If our healthcare providers cannot take care of these people now, what makes us think that we will provide for them in the future? We have to be prepared for the demographic changes that are occurring in our country. Some of these changes include things like increasing racial diversity, Latino/as projected to become the largest racial or ethnic minority in the US, an aging population, and Millennials becoming the largest adult group generation. Though we are considered one of the most developed countries globally, we are still at the barrel's bottom in terms of health. It is time to revolutionize how we look at health and how we serve others' healthcare needs. For example, instead of making patients meet their providers, we should equally encourage our providers to meet where their patients are as well.

One way health professionals can meet their patients is outside the health exam room is by being involved in designing programs and solutions that will help eliminate health inequities in our country using the SDOH framework. For example, a proper health intervention would constitute a program that offers subsidized housing, building a better high-school-to-college pipeline, and increasing job security for those who have lost their businesses and jobs. Focusing on fundamental causes will allow us to stage a proper intervention because it would address many social determinants of health and improve overall health for the entire community. After fundamental causes are addressed, the focus can then shift toward proximal risk factors. We can offer community education courses and design health campaigns to improve behavioral and psychological perspectives on health and teach community members how to take care of each other.

Another intervention for the neighborhood could be giving residents better access to alternative transportation that will provide more avenues for physical activity, better the air quality because of reduced pollution, and allow families to spend their money elsewhere instead of on a vehicle. One way to do this is to build more sidewalks with bike paths and more public transportation that is cheap and accessible. Suppose any of the issues above are addressed. In that case, they will create better health outcomes along a more extended timeline because these structural changes will bring the community closer together and positively impact their future generations. Several of these different interventions will be expanded on in parts two and three of this book.

REFERENCES

1. Braveman P, Gottlieb L. The social determinants of health: it's time to consider the causes of the causes. Public Health Rep. 2014;129 Suppl 2(Suppl 2):19-31.

2. Flaskerud JH, DeLilly CR. Social determinants of health status. Issues Ment Health Nurs. 2012;33(7):494-497.

3. Lozano R, Naghavi M, Foreman K, et al. Global and regional mortality from 235 causes of death for 20 age groups in 1990 and 2010: a systematic analysis for the Global Burden of Disease Study 2010. Lancet (London, England). 2012;380(9859):2095-2128.

4. Solomon LS, Kanter MH. Health Care Steps Up to Social Determinants of Health: Current Context. Perm J. 2018;22:18-139.
5. McEwen BS. Neurobiological and Systemic Effects of Chronic Stress. Chronic Stress (Thousand Oaks). 2017;1.
6. Juster RP, McEwen BS, Lupien SJ. Allostatic load biomarkers of chronic stress and impact on health and cognition. Neurosci Biobehav Rev. 2010;35(1):2-16.
7. Beyer KM, Zhou Y, Matthews K, Bemanian A, Laud PW, Nattinger AB. New spatially continuous indices of redlining and racial bias in mortgage lending: links to survival after breast cancer diagnosis and implications for health disparities research. Health Place. 2016;40:34-43.
8. Mendez DD, Hogan VK, Culhane J. Institutional racism and pregnancy health: using Home Mortgage Disclosure act data to develop an index for Mortgage discrimination at the community level. Public Health Rep. 2011;126 Suppl 3(Suppl 3):102-114.
9. Diez Roux AV, Mair C. Neighborhoods and health. Ann N Y Acad Sci. 2010;1186:125-145.
10. Palumbo AJ, Wiebe DJ, Kassam-Adams N, Richmond TS. Neighborhood Environment and Health of Injured Urban Black Men. J Racial Ethn Health Disparities. 2019;6(6):1068-1077.
11. Liu SR, Kia-Keating M, Santacrose DE, Modir S. Linking profiles of neighborhood elements to health and related outcomes among children across the United States. Health Place. 2018;53:203-209.
12. Foraker RE, Bush C, Greiner MA, et al. Distribution of Cardiovascular Health by Individual- and Neighborhood-Level Socioeconomic Status: Findings From the Jackson Heart Study. Glob Heart. 2019;14(3):241-250.
13. Penman-Aguilar A, Carter M, Snead MC, Kourtis AP. Socioeconomic disadvantage as a social determinant of teen

childbearing in the U.S. Public Health Rep. 2013;128 Suppl 1(Suppl 1):5-22.

14. Owens DC, Fett SM. Black Maternal and Infant Health: Historical Legacies of Slavery. Am J Public Health. 2019;109(10):1342-1345.

15. Ozimek JA, Kilpatrick SJ. Maternal Mortality in the Twenty-First Century. Obstet Gynecol Clin North Am. 2018;45(2):175-186.

16. Khreis H, Kelly C, Tate J, Parslow R, Lucas K, Nieuwenhuijsen M. Exposure to traffic-related air pollution and risk of development of childhood asthma: A systematic review and meta-analysis. Environ Int. 2017;100:1-31.

17. Han L, You D, Gao X, et al. Unintentional injuries and violence among adolescents aged 12-15 years in 68 low-income and middle-income countries: a secondary analysis of data from the Global School-Based Student Health Survey. Lancet Child Adolesc Health. 2019;3(9):616-626.

4

Power, Wellness, and Identity

Historical abuse of power and how it relates to modern-day health and identity formation

> *"Let no one ever intimidate you, you are standing on no one's ground. But again, some have claimed the earth as their own and usurped power from the rest of us. But they are usurpers; power belongs to every one of us."*
>
> —*Bangambiki Habyarimana*

Identity Intersectionality

★ ★ ★

During my early educational career, I was privileged enough to be able to learn about the importance of identity and power and how it relates to wellness. All the connections started to come together when I was sitting at a required workshop for one of my classes that talked about learning the importance of identity. I was thrilled to know that this was a recurrent theme in my education, however I was able to notice that not everyone was as thrilled as I was. One specific example was when I was in one of my master's degree class where we got to learn about racial identity and health disparities in the context of environmental health. I was captivated by the lecture and as such I was not really paying attention to what others were doing. It was evident that many other people were not paying attention because I overhead my classmates in the bathroom saying how bored they were. One classmate said, "This is so stupid, I don't get why we need to learn about this. It is irrelevant and not useful. I am going to leave a bad review." Everyone else started shouting in agreement and also laughing at our speaker. They were unaware I was in the bathroom as I was in the furthest stall from the entrance. However, as soon I walked out to wash my hands, they all remained silent. Unsurprisingly, my classmate was a white male that denied the existence of privilege in our previous class discussions. What was surprising, is that this was a group of guys who were pursuing a career in the health field and yet failed to acknowledge the importance of power and identity and the influence it had on wellness.

Learning about your own identity and its intersections with power and health is crucial for First-Generation Americans (FGAs) because it allows us to appreciate the patterns of human knowledge, belief, and behavior. Knowing your identity is important to generate knowledge to transmit our stories to the next generation. By being aware of your identity, you can maintain your cultural heritage and pass it on to your children. When talking about identity intersectionality, we must also consider how different identities intersect with each other. For example, I identify as a Mexican American man who is a college student. In this statement, I've highlighted a few identities: 1) my nationality; 2) my gender identity, and 3) my identity as a student. These identities exist with each other and are essential to understand as intersections when your life begins to interact with different power dynamics in various social settings.

Identity intersectionality is about the various interactions among different identities rather than just considering each individual identity in silo.[1] It gets more complicated when we start to talk about easily visible identities compared to others that can't be seen. For example, by looking at me, you can assume that I am a Brown person of color. But just by looking at me, you can't tell that I also identify as a physician scientist. Other examples of identities that are easily visible include race or assumed gender.

In contrast, negative imposed identities tend to be internalized, or they exist at the systemic level, and they are not that easy to identify between diverse people. When we cannot affirm and reconcile the intersections of people's identity, power dynamics, and historical context, we cannot have essential discussions because we will lack perspective. Therefore, members of any group who are conveyed privilege by birth or perceived by others as having privilege and who discriminate, hate, exploit, or reap unfair advantages over groups with less power should be scrutinized.

The relationship between power and health has to be understood at the intersectionality of multiple identities, including age, gender, socioeconomic status, race or ethnicity, sexual orientation, and religious and spiritual dogmas. To best learn this relationship between power and health, there must be prolific discourse around diversity and identity issues concerning racial relations both in individual and communal contexts. It is essential to understand the historical connection between health and politics of identity. We aim to build an all-embracing self-awareness as FGAs further in our quest to understand power and privilege dynamics. We must also explore our assumptions of identity and how other people identify. Only then will we recognize the matters that affect everyone without any personal judgment or bias toward others. We must challenge our perspectives, values, goals, and practices that distinguish us as humans or as institutions.

Current American culture seeks to assimilate others' culture, language, and history into the dominant culture, where diverse cultures are misrepresented and eradicated overtime to preserve America's whiteness.[2] Whether through discrimination, oppression, exploitation, or marginalization, this abuse of power can impact the feelings, beliefs, and values of an individual or change their actions, behaviors, and language. At the systemic level, power will influence the legal system, educational system, public policy, hiring practices, mainstream media. The influence ultimately sets our societal and cultural norms and the collective idea of what is considered fair. Our experiences in diverse aspects are negatively impacted by these mechanisms of domination and subordination, primarily our well-being.

Power and Privilege

★ ★ ★

In its more simplistic form, power comes down to having the ability to influence decisions and behaviors that impact people in their environment. Power can come in many different ways, including profession, money, education, and politics. By having power, people can advocate for themselves and influence decisions that can improve their access to health care, clean air, better working conditions, and educational opportunities. In this way, power is related to health equity, and imbalances in power are the root cause of racial discrimination and health disparities. The fact that power has been explicitly and, more recently, implicitly denied to several communities is a problem that needs to be acknowledged and addressed. Therefore, as FGAs, we must make it a priority to support communities to build social connection and power, particularly for people of color, for those in the low socioeconomic strata, especially for those who have been directly impacted by poverty, systemic racism, and exploitation. As a country, we need to recognize that the imbalances with abuse of power cannot be resolved by adopting a status quo mindset similar to what we have now; instead, we must reframe our understanding and pursue equity. Increasing power capacity in our communities can improve our ability to promote sustained and systemic change and promote transparency, accountability, and fairness in the most utmost important aspects of our lives. In doing so, we can start to address the health inequities that plague our society and build a brighter and healthier future for all.

If an imbalance in power is the underlying cause of health inequities in our society, then the abuse of power or the misuse of privilege is how we got there in the first place. Privilege is an entitlement that grants a specific group the benefit of immunity and advantages that are not extended elsewhere. Privilege can manifest itself in many ways, including but not limited to skin color, education status, job title, etc. Harmful privilege occurs due to an unchecked abuse of power that becomes a prominent force that creates systemic oppression. Since the use of institutional privilege can oppress, it is usually at the expense of others without power.

While acknowledging that the misuse of power is the root cause of several inequities, at the same time, we need to recognize that power can also put us in a position to cultivate positive change. We must move forward united and strive to create a world where power is no longer abused and misused by people. As a training physician, I have committed my life and taken an oath not to harm. Traditionally, this refers to not hurting individuals who seek out medical care. However, as mentioned in earlier chapters, emerging evidence in the health sciences has demonstrated that power has several different dimensions and can cause systemic harm in other ways.[3] Mainly through the passing of laws, issuing orders and decisions, and setting up an agenda that selectively chooses which issues need to be discussed and addressed or ignored.

The promotion of equity in power and health depends on improving the social, economic, and cultural conditions that shape our health outcomes. As we continue to learn more about the human condition and experience, we are only starting to uncover health intricacies. To some, being healthy simply means not being sick. However, I have learned that health means more than just not being sick, and it encompasses the intersections of biopsychosocial elements, including spiritual well-being and emotional well-being. Understanding power dynamics and identity is essential when examining the historical course of health outcomes in certain communities, such as the Native Americans in the US, because it allows us to explore the root causes of how we got to where we are.[4] By having an adequate understanding of identity, power, and privilege, we can apply this framework to other populations, which will provide us with a new perspective in appreciating and investigating the course of health outcomes in the country.

Historical Abuses of Power, Intergenerational Trauma, and Epigenetics

★ ★ ★

An intergenerational understanding of racial health disparities among people of color is an emerging topic that needs more attention because it has highlighted several risk factors that result in health disparities. Since the initial moments of colonization, repeated attacks on individuals' public health have resulted in different types of trauma and post-traumatic effects. These effects that have created racial health disparities in America stem from physical, emotional, and psychological traumatization across the entire continent over the past several hundred years.[5,6] In this context, colonization means the action or process of settling among and establishing control over the indigenous people of an area. As previously mentioned, we have even started to learn and will also highlight in the next chapter that repeated traumatization can cause biological changes that influence your probability of getting a disease more so than others.[7] These traumas have not only impacted our ancestors, but in several cases, they have also affected us by directly modifying our DNA over time.

In the emerging field of epigenetics, negative experiences over multiple generations negatively influence the next generation's health.

Epigenetics refers to the study of how your behaviors and environment can cause changes that affect the way your genes work. This mechanism that causes poor outcomes is also known as historical trauma. Our country's abuses of power have created negative multigenerational health impacts in communities through the life course and early adverse childhood experiences, which will be highlighted in more detail in the next chapter. Essentially, we need to know that there is a biological integration of adversity during sensitive periods like childhood that can result in mental health issues and eventually impact future generations.

As we have learned in previous chapters, there is no biological basis for race because we all have the same DNA as humans. However, different genes express different patterns. The difference in expression can be expected; however, it can also be pathological or disease-causing. For example, epigenetic mechanisms such as DNA methylation and histone modification are inherited through cell divisions over multiple generations.[8] These epigenetic regulation mechanisms work as either off or on. That is, you are either expressing the gene or not. More importantly, we know that epigenetic mechanisms are also interconvertible, meaning that they can be reversed. Unusual gene expression through epigenetics causes disease, but there is a therapeutic possibility because of reverting epigenetic changes.

In other words, a person's health starts before they are born and is affected by the mother's health long before she is even pregnant, for example, because of the stress from historical trauma, which can change someone's health negatively if it is chronic. And as we have learned in redlining, disadvantages through the constant marginalization of neighborhood economic status happens over time and is commonly transmitted to the next generation. As a result, a multigenerational perspective is crucial to understanding the relationship between neighborhood environments and their health effects on later generations.

Through the comprehensive study of the intersections of identity, power, and health, we can better predict disease in individuals across generations and are better equipped with how to advocate for people. We also have a better understanding of their narratives and stories, which I will be telling more of in part two of this book. Now that we have a solid foundation in several of these topics, we can better begin to understand the different racial health disparities that plague our society.

REFERENCES

1. Byrd WC, Brunn-Bevel RJ, Ovink SM. Intersectionality and higher education: identity and inequality on college campuses. New Brunswick: Rutgers University Press; 2019.
2. Telles EE, Ortiz V. Generations of exclusion: Mexican Americans, assimilation, and race. New York: Russell Sage Foundation; 2008.
3. Farmer P. Pathologies of power: health, human rights, and the new war on the poor : with a new preface by the author. Vol 4. Pbk. ed. Berkeley: University of California Press; 2005.
4. O'Neill L, Fraser T, Kitchenham A, McDonald V. Hidden Burdens: a Review of Intergenerational, Historical and Complex Trauma, Implications for Indigenous Families. J Child Adolesc Trauma. 2018;11(2):173-186.
5. Phillips-Beck W, Sinclair S, Campbell R, et al. Early-life origins of disparities in chronic diseases among Indigenous youth: pathways to recovering health disparities from intergenerational trauma. J Dev Orig Health Dis. 2019;10(1):115-122.
6. Conching AKS, Thayer Z. Biological pathways for historical trauma to affect health: A conceptual model focusing on epigenetic modifications. Soc Sci Med. 2019;230:74-82.

7. Heinzelmann M, Gill J. Epigenetic Mechanisms Shape the Biological Response to Trauma and Risk for PTSD: A Critical Review. Nurs Res Pract. 2013;2013:417010.

8. Bianco-Miotto T, Craig JM, Gasser YP, van Dijk SJ, Ozanne SE. Epigenetics and DOHaD: from basics to birth and beyond. J Dev Orig Health Dis. 2017;8(5):513-519.

The Racial Health Inequities that Exist in America

An overview of the different institutionalized inequities in the US

> *"Of all the injustices in the world, injustice in health care is the most shocking and inhumane."*
>
> —*Dr. Martin Luther King, Jr.*

The Causes and Basics of Health Disparities

★ ★ ★

During one of my internal medicine rotations, I remember I had the opportunity take care of a thirty-year old patient who was a Black female. In the morning before starting to round on patients, the cross-cover team was giving the hand-off to us and explained to us that she was in the hospital because she had end-stage kidney disease from her uncontrolled diabetes and as a result of her failing kidney, she also started to have heart failure. I was completely perplexed as to how someone so young can already have heart and kidney failure? I immediately thought it might have to do because she must have had some genetic predisposition, as we don't tend to see those diseases in that severity at such a young age, but it turned out that the reason why she was already on her death bed is because she did not have any insurance to buy her medications. At the end of the handover, the overnight resident said, "Oh by the way, she is also a really angry lady and is hard to work with." At this time, I realized that she was being stereotyped as the angry Black lady and that her concerns are probably not being heard out. I immediately realized that I needed to see this patient and just hear her out and see if I could address her concerns. She shared that she does not trust her doctors because one doctor told her that she could not qualify for a new kidney and that because of that he was going to try to make her kidney worse. She shared that she is in pain and that no one is doing anything about her pain. She was unaware of the severity of her condition and insisted that she was, "fine." Unfortunately, even after trying to build a therapeutic alliance, I failed to try to understand how I can help make her condition better. I felt hopeless and overwhelmed of this massive health disparity that I was seeing right before my eyes. This was the first time during a hospital encounter when I was able to see the direct impact of racial health inequities that exists in America.

Health inequities in America are rooted in the complex interactions between people and institutions through socioecological models that keep some humans at the margins of life. One way to understand how these complex interactions work with each other is by highlighting how health disparities impact each racial/ethnic group. This chapter will be looking at different topics such as geopolitics, built environment, socioeconomic indicators, and access to medical care to see how and where these health disparities manifest in society. After we have a better overview of what health inequities are and where they stem from, we will discuss specific types of disparities for each racial minority group in America. This chapter aims to critically analyze which processes in the environment can initiate and sustain racial disparities in health over generations.

Over time, I have learned about different causal pathways and patterns that can lead to mortality and morbidity. Morbidity can refer to different disease states, while mortality refers to death. From learning it in the classroom through my studies to seeing it in my life as in my grandmother and father's case as you will soon learn, several casual pathways are at play instead of just one singular cause. Using race without contextual information for health outcomes is limited and fails to acknowledge the underlying cause of diseases. As we currently know, heart disease is the leading killer in the world. Mortality trends from heart conditions are close to three times higher among low-income African Americans and white Americans than among their middle-income counterparts.[1] What this means is that if we were to have two people, one white and one Black, who have the same job, make the same amount of money, and have the same amount of education, Black Americans are still a lot more likely to have adverse outcomes compared to their white counterparts.

How can people, fellow humans, have differences in mortality based solely on the indicator of race? As we discussed in chapter two, in medical education, we learn that race is a risk factor for increased mortality and morbidity for several diseases. But why is this the case if race is socially constructed and not biological? The biological weaponization of race has caused inequities in health and socioeconomic opportunities in America. In other words, as we have also learned, it is racism that causes adverse outcomes, not race.[2] Institutionalized inequities stem from colonial and imperial historical forces that have shaped the world we still live in today. The etiology of racial health disparities is not biological. Instead, they result from a system designed not to consider everyone's progression in the human species. Institutionalized inequities have resulted in significant differences in opportunities in life. If we allow our institutions to continue to uphold these values, America will continue to prosper from exploiting others and our natural earth resources.

For several hundred years, non-white humans who were stripped of power did not have access to opportunities to control their lives and live a prosperous life.[3,4] The result of different power dynamics and exploitation of those power dynamics has resulted in a society that believes one race is superior to others. Differences in power have resulted in geopolitical and financial disputes, resulting in harming fellow humans outside of the dominant group's social sphere, consequently leading to a cycle of poor health and low socioeconomic status or marginalization. Not only do people who have less power have less access to fundamental human health rights, but there is also a decreased perceived risk of health risk factors and the reduced perception of having the ability to control their lives. Most often, the decreased perceived danger of these different risk factors is a result of a lack of access to education and cultural norms.

After hundreds of years of living in a racial, social construct, racial minorities will either not have access to vital health rights or fail to ask for care due to the lack of trust in systems. The mistrust among racial minorities in US systems originates with the various historical traumas they have faced. Minorities face daily discrimination in the forms of direct racism and indirect racism, such as racial microaggressions. Microaggressions are commonly understood as brief, everyday exchanges that sends unconscious racist messages to people of color because they belong to a specific racial group. One example of a microaggression I get a lot is when I speak to people in English. When they say things like, "you speak really good English," or say "you are not like the rest of them," it implies that Mexican's don't speak good English and that they are not high achievers.

Microaggressions themselves are not direct causes of racial health disparities, but they put people at a higher risk of developing the disease because of the chronic stress that people are under all the time due to these microaggressions. The entire process puts minorities at a higher risk of adverse health outcomes and racial health disparities.[5] We have learned that it is more challenging to defend yourself from disease when your body is under chronic stress. Some of these expected health disparity outcomes include posttraumatic stress disorder, traumatic brain injuries, anxiety, self-destructive behavior, excessive use of substances as coping mechanisms, and lastly, not being able to take care of their medical concerns due to cultural barriers.[6] These different outcomes can have unique pathways that lead to them; however, they are all connected because they all stem from the same social environment.

One of the disproportionate disadvantages that minority groups face is unethical medical treatment through institutionalized oppression. For example, Black people are less likely to get referred to special treatment, less likely to get prescribed pain medication, and less likely to even get standard therapy due to providers' bias and how current health insurance is set up.[7] When health providers are clouded by bias when they interact with their patients, it can result in medical, cultural clashes. This scenario is especially so when we talk about culture-bound syndromes, which are diseases that are bound to a specific culture.

Culture Bound Syndromes

★ ★ ★

In medical anthropology, a culture-bound syndrome is a combination of psychological and physical symptoms that are recognizable within a specific society or culture. Several of these syndromes originate from spiritual, evil, ancestral spells that cause patients to present in the clinical space.[8] Some culture bound syndromes in the Latino/a community include things like *susto, empacho, nervios, mal de ojo,* and *nervios.* To further describe one of these syndromes, I will elaborate on *empacho* which is usually used to describe the symptoms of constipation, stomach aches, nausea, vomiting indicating that there is a potential obstruction of the stomach or intestine. I remember when my mother would tell me not to eat raw cookie dough because it would make me get *empachado.* When patients go to see doctors for these complaints, it can often be dismissed. However, if you go to a predominant Hispanic community and you talk about these conditions, they will know what you are talking about.

As we can imagine, in medicine, everything is so objective that physicians cannot appreciate the different possibilities that can cause disease in the patient population. Considering that there are often no biological explanations for these diseases, health providers may dismiss the complaint or even give the wrong diagnosis. Rather than just dismissing these complaints from diverse patients from different cultures, we can instead focus on learning more about their condition and supporting them the best way we can.

One tool available to us as clinicians and workers of health are the Arthur Kleinman questions, which help us understand where the patient or person seeking care is coming from.[9] Some examples of these questions include things like What do you think caused your problem? What do you think your sickness does to you? What do you call your problem? What do you fear about your sickness? These questions are crucial to better understanding the culture bound syndrome of the patient, but also important because it puts a focus on what the patient thinks rather than the assumption of that doctors know everything. Being aware of these various culture-bound syndromes based on the populations you work with is essential to building a therapeutic alliance with patients and to also aid in moving us towards a society that can eliminate disparities in health.

When marginalized throughout the process, these patients are at a higher risk of falling into a vicious cycle of neglected healthcare needs because their healthcare providers are not culturally responsive. As a result of not accessing care that addresses the patient's complaint, their health continues to decline. It puts them at a greater risk of not being healthy enough to work, contributing to poverty cycles. In addition to facing culture-bound syndromes, members in such communities may, at the same time, also have higher incidence and prevalence rates of preventable adverse health outcomes. When patients come in to talk about a culture-bound syndrome, and it is dismissed, they are less likely to want to engage in any other preventive care measures that should be encouraged.

This misunderstanding of values and beliefs between providers and patients perpetuates unethical medical practices, resulting in poverty and poor patient health. As a result of these clashes, psychosocial stress will deteriorate their health defenses. Psychosocial stress factors are sometimes hard to see, and they sound like an abstract idea. In the next section, I will highlight how psychosocial stress factors can present themselves from the perspective of migration, which is just one of the many stressor's patients may face. Can you think of any psychosocial stress factors in your life and how they might be related to your health? What can you start doing to be more aware of these stress factors and find ways to reduce them?

The Myth of the Model Minority for Asian Americans

★ ★ ★

The model minority is one of the core ideologies that continue to cause oppression and other forms of racism among Asian Americans and other minority groups. The model minority is an idea that members from a specific racial group, in this case, Asian Americans, are perceived to have a higher degree of success than other minority groups as such should serve as a reference for everyone else to emulate.[10] It is a false narrative created by white supremacy to highlight what is perceived as the successful assimilation of Asian Americans into mainstream culture. It is weaponized against other racial minorities who are then punished for not achieving the same standard.

The standard narrative promoted by white supremacy is that Asian Americans as a minority group are smart, polite, successful, law-abiding, hardworking, thriving, and model citizens.[10,11] As such, they are a prime example of how other minorities, too, can "overcome" racism and discrimination to live the American Dream. This narrative is a profoundly harmful myth because it's been designed to maintain a specific racial hierarchy and power dynamic. First, it pits minority groups against each other. For example, take the common saying, "Why can't Latinos be more like Asians when it comes to being high-achievers in school?" or, "If Asian minority groups can achieve success, why can't Black people do it, too?" Such questions manufacture racial conflict among minorities. Besides, taking the white man out of the racial discourse and putting the onus on each other distracts minorities from the real culprit—the more extensive system and status quo crushing everybody.

Second, the model minority myth doesn't account for multiple counternarratives within the Asian American community. By overgeneralizing, it reduces the diversity of the Asian community to a stereotype. The heterogeneity of the Asian community is vast and includes roots that can be traced to over 20 countries in East, Southeast Asia and the Indian subcontinent – all having unique histories, cultures, languages and several other characteristics.[10] On top of creating a stereotype, the myth pressures Asians to assimilate and be "whiter," keeping them stuck in subservience to white supremacy.[10,11] We have learned the dangers of assimilation when we should instead be working toward acculturation of different groups into American society. The combination of the model minority myth and white supremacist attempts to erase the identity of the diversity of people misrepresents, distorts, and simplifies the reality of American society. The truth is that there is no model minority—all minorities are unique, and this is what allows us to be great.

The model minority myth has a significant impact on Asian communities, First-Generation, immigrant, refugee, or all of the above. But while it has a common origin and implications across all Asian groups, it disproportionally affects Asian refugee groups more than other Asian immigrant groups. It has more consequences in Asian refugee groups because they usually come from a lower socioeconomic status. However, if we look at socioeconomic data of Asian groups as a whole, we see that Asians are some of the top earners in our country. Therefore, this apparent "lack" of need in the overall community diverts resources away from underserved Asian groups. Some of these resources include Medicaid, affirmative action, and other benefits that allow organizations to engage in a healthy lifestyle and future. One specific example is Southeast Asian American refugees, such as the Hmong, who are frequently damaged by the model minority stereotype.[12] The Hmong are just one of the many groups who have to undergo this daily battle.

The case of a Hmong child and her medical doctors documented by Anne Fadiman illustrates how a traditionally disadvantaged minority group can clash with their healthcare providers.[13] In the case of Lia, the main child of the story, she was diagnosed with epilepsy. But in the Hmong culture, someone with epilepsy is considered to have a gift; it is believed that having epilepsy means you are a chosen one and have great honor. The medical professionals strongly opposed this cultural belief. They decided to explain to the family that it was not a spiritual gift but rather a life-threatening illness if it got severe. The medical professionals completely dismissed the patient's family's beliefs and assumed the patient's needs without accommodating cultural differences and proceeded with what they thought the patient needed. In other words, healthcare professionals were not culturally responsive to their patients. Rather than taking the time to understand their patient and cultural background, the team decided to treat the medical condition instead rather than treat the patient. What I mean by this is that they were more focused on treating epilepsy that they were unable to treat Lia as a whole, especially concerning how epilepsy is perceived in Hmong culture. It came down to a tug of war between the patient's family and the healthcare staff, with the patient's life on the line.

What could they have done instead? The team could have attempted to form a therapeutic alliance with their patient by demonstrating compassion and a mutual understanding of what was important to their patient. In this case, healthcare providers could have demonstrated cultural responsiveness by taking the time to understand the condition in their particular culture and working with the family to ensure that the patient was safe and healthy. Had the medical team been culturally responsive and acknowledged the role of the condition in Hmong beliefs, there could have been an opportunity to talk about all the appropriate treatment options. Rather than trying to make it seem like they are curing a "disease," they could have taken the approach of treating the patient as a whole, which would have been much more effective. With our country's demographics changing exponentially, healthcare professionals must keep up with the skills and abilities needed to deliver care to their patients successfully.

A Historical Perspective of Health Inequities among Black and African Americans

★ ★ ★

The African diaspora, which includes Black and African Americans, has seen countless documented atrocious human rights violations against them that have contributed to much of the health inequity we see in the community today.[14] These atrocities range from as early on as J Marion Sims doing surgery on African slaves without anesthetics to perfect his medical procedures for white women in the 1800s.[15] Or the Tuskegee Syphilis Experiment, dermatologic medical experimentation on prison populations, and even today with experiments being conducted on communities in low-income countries. Throughout history, we have seen Black people's repeated traumatization through forced removal, slavery, and inhumane medical experimentation.[14] Considering that Black people's plight in our country is so unique, it isn't easy to choose one case representing everything they have had to face. However, one case I do want to highlight and that I believe can illustrate several of the barriers that Black people face in health care is the story of Henrietta Lacks's exploitation by the medical system.

Henrietta was born in 1920 and was diagnosed with cervical cancer in 1951. The appearance of the tumor was unlike anything that had ever been seen by the examining doctor. The doctors removed the cells from the cancerous tumor for research purposes without her knowledge or permission, which at the time was standard procedure. In the medical ethics community, Henrietta is now known as the African American woman who was the unwitting donor of cells used to create an immortal cell line for medical research known as the HeLa cell line.[16] Although she died from cancer that same year, her cells continued to live on in medical research. They were sold and used for many experiments and put into mass production, all without her children's knowledge. In the early 1970s, the family had barely begun to learn about how her doctors removed her cells.

Her cells became privatized in 1951, and they created a massive industry for research to the point that doctors even mailed them to scientists around the globe. Today, Henrietta's cells have been used in countless scientific pursuits. Some of the experiments conducted on her cells include trying out different types of cancer drugs, studying the effects of radiation and toxic substances, testing human sensitivity to tape glue cosmetics and other products. Even gene mapping has raised several questions today regarding the invasion of her family's privacy since every scientist can now get access to the Lacks family's genome map. Since the first removal, scientists have grown some twenty tons of her cells. Despite the increased attention, in this case, Henrietta's family has never been compensated. They have had to grow up in poverty and lack health insurance to even pay for medical care.

The case of Henrietta Lacks is just one tiny glimpse into the health inequities that Black people face both in terms of systemic exploitation, as we saw with Henrietta, to injustices that are more visible, such as the mass killings of unarmed Black men. The overt murder of Black men in America has its foundations since slavery. Over time, instead of being eliminated from society, how America kills its

Black men has just transformed, notably through the modern-day criminal justice system.[17] The criminal justice system has also had significant health impacts on Black women, such as through the similar mechanism of mass incarceration. Additionally, one of the less visible ways that Black women experience health inequities is through high rates of maternal and fetal morbidity and mortality, or in other words, the high rates of mothers and their babies dying and experiencing disease in the population.

Health disparities among Black women have been well documented throughout history, yet it continues to be a prominent problem in modern America. The disparities of primary health indicators such as maternal/fetal death and morbidity rates are among the most staggering differences we see.[18] It is staggering in the sense that the disparities differ substantially compared to their white counterparts, even after controlling for variables like income, education, age, etc. Meaning that even if everything else is equal, you are significantly more likely to experience higher maternal and fetal complications if you fall under the Black racial category. For years, people have been asking why this is the case, as there is no biological basis for race. As we have learned, the sociopsychological factor of racism plays a part. While it is not the only explanation, it is undoubtedly a root factor in these differences we see.

A Personal Case Study of Migration-Related Psychosocial Stress Factors

★ ★ ★

In my life, I have faced several psychosocial stressors. As a First-Generation American, childhood adversity was no stranger in my life. For example, I could not have my parents participate in several aspects of my childhood outside of the home because of fear of deportation. Legal status in the US plays a crucial role in how people live, for example, living in fear and not accessing healthcare rights.

As an FGA, I had the privilege not to face detention and deportation, which several of my family and friends had to go through. Family separation as a result of removal also takes a toll on emotional and psychological health. And when my family members became sick, we would have financial stress from a lack of insurance access. History keeps repeating itself. This example is what it feels like to have a traumatizing emotional experience.

I have witnessed how these factors have intensified over several decades due to racist rhetoric and increased anti-immigrant sentiment. Trauma, fear, depression, loneliness, sadness, stress—all of them combine into a cocktail of psychosocial stressors. All because I am considered an "alien" in the land of my ancestors. My family's migration-related psychosocial stressors profoundly impacted the emotional experiences in my life and others around me. Some primary structural vulnerabilities that cause personal hardship among immigrants are pre-migration exposures and adversity, deadly boundary intersections, detention and deportation, undocumented citizenship status, family separation, and extreme poverty.[19]

Is my life an illustration of migration-related psychosocial stress factors associated with First-Generation American immigrants? I would argue, yes. FGAs undergo unique psychosocial stressors that result from migration and forced removal that negatively impacts health. As a result of structural policies limiting our ability to acculturate into the US, we also disproportionately face psychosocial stressors related to being First-Generation in the US, mostly affecting our mental health. Given the heavy emotional toll of migration and the direct impact that regional legislation and border security has had on my family's well-being, I argue that these structural pathologies are an essential mechanism for health inequities in the indigenous community. Science has demonstrated that social and economic inequality poorly affect overall health. One of the most basic examples in which inequities impact the body is adverse neurodevelopmental health, which carries known psychological and pathophysiological responses.

It is long overdue for people who are FGAs to have an equitable path to citizenship. Additionally, our generation must also focus on developing policies that make it easier for racial minorities to obtain health services and legal services. Health disparities are a threat to overall public health because they take a toll on healthcare spending and limit everyone's opportunities to live productive lives.

A Brief Description of Inequitable Health Pathways among American Immigrants and Native Americans

★ ★ ★

Health disparities that are understood in broader historical, geographic, sociocultural, economic, and political contexts are essential, especially for us to have a comprehensive understanding of the plight of Native Americans and American immigrants in the country. In most cases, American immigrants are Native Americans because they are native to the continent of America. For Natives, it is obvious how forced removal from your land can disrupt life in brutal ways. On top of facing forced migration, Natives have died from diseases, and flat-out genocide carried out by colonizers. A notion that needs more careful examination is how these actions performed several hundred years ago are continuing to impact the lives of their descendants. The consequences of the forced removal of Natives in the past are still relevant because it might explain some of the most severe health disparities we see among Native communities today.[20]

After Natives were concentrated on reservations, they had little access to healthy foods. This is still true today. Early in their forced removal, European settlers only provided bread and foods with poor nutritional content. For several hundred years, bread was one of the main foods that were consistent in their diet. We know today that eating only carbohydrates is bad for your health, and we are only starting to see how it impacts our current generation. If we examine the health status of the Native population, they have some of the worst outcomes out of all racial minority groups. Could it be that several hundred years of oppression is part of the equation of why we see the rates we see today? I think the lessons in this book show us that it certainly can.

On the other hand, Natives who have had to immigrate to the US from another part of America, or "Latinos" as we call them within the continent, have a different set of experiences that can lead to adverse health outcomes, most notably from the migration process. The migration process generally happens in three stages: pre-migration, migration, and post-migration/resettlement. The migration process has been well studied. It has been shown that when immigrants and refugees are put through various obstacles during each stage of displacement, it results in adverse health outcomes for the community. During pre-migration, the immigrant or refugee's health and socioeconomic status will largely influence their health outcomes during later stages of migration and resettlement in the US. Immigrants come to the US on their own will for multiple reasons, such as wanting to escape poverty to provide a better life for their children and pursuing educational opportunities they do not have access to in their home countries. On the other hand, refugees come because they are usually fleeing persecution in their home countries due to their race, ethnicity, religion, or multiple other identities they choose to embrace. Before migrating, immigrants and refugees are most likely exposed to trauma, war, loss, and psychological warfare, which lead to other adverse health outcomes such as mental illnesses like depression.

During migration, immigrants and refugees face countless traumas, as I have mentioned, for example, deadly boundary intersections, detention, and deportation, and family separation. Additionally, they have to deal with past injury, damage of possessions, loss of family members, and loss of financial and social capital that have followed them throughout their entire journey to the US. During resettlement, this post-migration stage continues to induce stress because it is difficult to assimilate into a new culture and re-learn how to survive.

Life in America for Native immigrants causes stress because of the preexisting hostile environment toward non-white-appearing people and the lack of necessities in such communities, as we have learned in previous chapters. Some of these necessities include spaces for physical exercise, access to healthy foods, and access to sanitary living conditions. When these necessities are not met, the likelihood of developing nutrient deficiencies, obesity, diabetes, and infections increase dramatically compared to the general population.

In this chapter, we have described some of the most fundamental racial health disparities in our country. Each one of these different health disparities has a unique story behind it and involves complex societal constructs that all lead to the same outcome of health inequity. As America's incoming physicians, it is imperative that we continue to scrutinize these current societal constructs and ideas that influence behavior and have been developed in a way that causes harm to people inside and outside of the hospital. With a solid understanding of the different health disparities that exists in our country and also a solid foundation in the concepts that explain these differences, we can now begin to translate many of these concepts into stories and practice in the next part of this book.

REFERENCES

1. Carnethon MR, Pu J, Howard G, et al. Cardiovascular Health in African Americans: A Scientific Statement From the American Heart Association. *Circulation.* 2017;136(21):e393-e423.
2. Bailey ZD, Krieger N, Agénor M, Graves J, Linos N, Bassett MT. Structural racism and health inequities in the USA: evidence and interventions. *Lancet (London, England).* 2017;389(10077):1453-1463.

3. Coates T-N. *Between the world and me.* New York: Spiegel & Grau; 2015.

4. DiAngelo RJ. *White fragility: why it's so hard for White people to talk about racism.* Boston: Beacon Press; 2018.

5. Slaughter-Acey JC, Sneed D, Parker L, Keith VM, Lee NL, Misra DP. Skin Tone Matters: Racial Microaggressions and Delayed Prenatal Care. *Am J Prev Med.* 2019;57(3):321-329.

6. Berger M, Sarnyai Z. "More than skin deep": stress neurobiology and mental health consequences of racial discrimination. *Stress.* 2015;18(1):1-10.

7. Del Pino S, Sánchez-Montoya SB, Guzmán JM, Mújica OJ, Gómez-Salgado J, Ruiz-Frutos C. Health Inequalities amongst People of African Descent in the Americas, 2005-2017: A Systematic Review of the Literature. *Int J Environ Res Public Health.* 2019;16(18).

8. Levine RE, Gaw AC. Culture-bound syndromes. *Psychiatr Clin North Am.* 1995;18(3):523-536.

9. Gaines AD. Culture, Medicine, Psychiatry and Wisdom: Honoring Arthur Kleinman. *Culture, medicine and psychiatry.* 2016;40(4):538-569.

10. Wu ED. *The color of success: Asian Americans and the origins of the model minority.* Princeton;Oxford;: Princeton University Press; 2014.

11. Yook EL. *Culture shock for Asians in U.S. academia: breaking the model minority myth.* Lanham: Lexington Books; 2013.

12. Lee S. More than "Model Minorities" or "Delinquents": A Look at Hmong American High School Students. *Harvard educational review.* 2001;71(3):505-529.

13. Fadiman A. *The spirit catches you and you fall down: a Hmong child, her American doctors, and the collision of two cultures.* 1st pbk. ed. New York: Farrar, Straus and Giroux; 1998.

14. Washington HA. *Medical apartheid: the dark history of medical experimentation on Black Americans from colonial times to the present.* New York: Harlem Moon; 2006.

15. Wailoo K. Historical Aspects of Race and Medicine: The Case of J. Marion Sims. *JAMA : the journal of the American Medical Association.* 2018;320(15):1529-1530.

16. Skloot R. *The immortal life of Henrietta Lacks.* 1st pbk. ed. New York: Broadway Paperbacks; 2011.

17. Alexander M. The New Jim Crow: Mass Incarceration in the Age of Colorblindness. *New Press.* 2010.

18. Owens DC, Fett SM. Black Maternal and Infant Health: Historical Legacies of Slavery. *Am J Public Health.* 2019;109(10):1342-1345.

19. Hynie M. The Social Determinants of Refugee Mental Health in the Post-Migration Context: A Critical Review. *Can J Psychiatry.* 2018;63(5):297-303.

20. O'Neill L, Fraser T, Kitchenham A, McDonald V. Hidden Burdens: a Review of Intergenerational, Historical and Complex Trauma, Implications for Indigenous Families. *J Child Adolesc Trauma.* 2018;11(2):173-186.

6

The Old vs. the New World

Historical perspectives from a Native Inmigrant

The earth was created with the assistance of the sun, and it should be left as it was. The country was made without lines of demarcation, and it is no man's business to divide it. I see the whites all over the country gaining wealth and see their desire to give us lands which are worthless. The earth and I are of one mind. The measure of the land and the measure of our bodies are the same. Say to us if you can say it, that you were sent by the Creative Power to talk to us. Perhaps you think the Creator sent you here to dispose of us as you see fit. If I thought you were sent by the Creator I might be induced to think you had a right to dispose of me. Do not misunderstand me but understand me fully with reference to my affection for the land. I never said the land was mine to do with as I chose. The one who has the right to dispose of it is the one who has created it. I claim a right to live on my land and accord you the privilege to live on yours."

—*Chief Joseph*

Immigration or Inmigration?

★ ★ ★

The first step in transforming a deeply embedded system in our society is by carefully studying the critical historical events that have put us in the position that we are today in twenty-first-century America. This idea is particularly true when we start talking about immigrants in political and historical discourse. When we talk about the "Old World," we usually refer to pre-Colombian times before the 1500s, while the "New World" is a term that was coined by European colonists when they were sailing across the ocean and claimed that they discovered a new land.[1] After massive European colonization across the world, America marginalized many of the Indigenous peoples across the continent for years. As time went by, there began to be more mixing between European colonialists and Indigenous Americans.

As a result of this historical crash, people in many Spanish-speaking countries across the continent are still native to those lands, although the government authorities do not formally recognize this fact all the time. While the people in power have historically neglected the sovereignty of several Indigenous groups across the world, this chapter highlights the story of Natives across the Americas to proclaim our sovereignty on this continent. And as such, appropriate repairs should be set forth to reverse the several hundred years of marginalization and exploitation of Native peoples in the Americas by European settlers.

The Mesoamerican region extends from the southwestern United States down to the border of Guatemala. Several of these areas, especially in Mexico and Guatemala, are known for their large ceremonial centers, mathematical expertise, intricate calendars, science, and astronomy.[2] What is the difference between immigration and inmigration in this context? Immigration means migrating to a new country to live, the common thing we think about when we talk about immigration. On the other hand, inmigration means the migrating of people within their ancestral land. In other words, several people who migrate from Mexico to the US or from the US to any other country in Mesoamerica are inmigrating within the land stolen from them. This idea is essential for the next generation to understand, especially as incoming FGAs from across the continent. We must be aware of our roots and also that we are still on our ancestral land. Not only will this give us a sense of belonging instead of a feeling of alienation in our land, but it will also give us a new perspective when we start demanding long overdue justice.

Knowing Your Historical Roots:
My Mexican-American Identity
★ ★ ★

One way you can start to learn more about your roots is by learning more about your family. Sometimes, information from your family is limited. Thus, it is essential to rely on secondary sources, including marriage records, migration records, and other historical documents found in several online formats, including Ancestry, 23andMe, etc. If you do not feel comfortable sharing your information with these resources, you can still learn about your history and conduct research on your background from your local library or even an online database. Knowing about your roots and ancestors is powerful, and I believe it is an important thing that we have to teach the next generation.

My roots are from Guerrero, Mexico, which is composed of the south-central region that has been home to many Native Americans, including the Aztecs, Maya, Inca, and several others.[3] Guerrero has evidence of human activity dating back to at least 300 BC, where the Olmec people inhabited the central and southern areas and have continued their influence on modern-day Guerrero. Anthropologists and archeologists suggest that the Olmec were one of the early builders of ancient America's first civilization. It wasn't until the tenth century when temples and pyramids were built, and the Aztecs ruled the region that consists of modern-day Guerrero, split into seven entities. However, Acapulco's region specifically never came under the control of the Aztecs but instead remained subject to local chiefs from the Tarasca, Mixteca, Zapotec, and Azteca civilizations.[4] This meant Acapulco was the central location where many of the leaders from several tribes came together. The fierce warrior-culture in this area built a rich legacy of art, science, and survival into the land.

Spanish, English, and French invaders murdered and forcefully displaced Native people around the American continent throughout history.[5] Throughout the fifteenth century, Acapulco became the capital for trade between Peru and Asia. Importing stolen slaves from Africa and parts of Asia had long been a common Spanish practice, and during the following century, Acapulco became the center for the slave trade.[4] Modern-day descendants of African slaves still live on the southern Pacific coasts. With Spanish conquest and the forced migration of African slaves, a mestizo population grew with immigrants from England, Portugal, and France until the early nineteenth century, though it remained heavily indigenous. Up until the nineteenth century, Mesoamerica was filled with revolution, from creating new rights for women in the Mexican Republic to gaining independence from Spain. After the bloody revolution that spanned the early decades of the twentieth century, the area prospered. Still, wealth was never shared equitably, so many began to migrate back to their ancestral homeland of the US, or, as our ancestors know it, Aztlán. Aztlán is the legendary and original homeland of the Aztecs, before leaving it to find a new home in the valley of Mexico. There is much speculation where Aztlán was located. Some say it spanned northern Mexico, which today is formally recognized as the southwest of the US including states like Texas, New Mexico, Arizona, California, Nevada, Utah, and Colorado.[6]

After several thousands of years of rebellion and migration in Mesoamerica, the ethnic group, presently known as Latino/a/x and Hispanic, was conceived. In America, we strictly had four races—Asian, Black, White, and Native—to creating the ethnic group of Hispanics or Latino/a/x. As we have learned in Part 1 of this book, the primary difference between race and ethnicity is that race is the fact or state of belonging to a racial, social group usually self-identified. However, the way others perceive your race is the other part of the equation. At the same time, ethnicity is the fact or state of belonging to a social group that has a common national or cultural tradition. As a result, Latinx/Hispanic people can consist of Black, Native, Asian, or white people. In the US, Latinx/Hispanics' heritage stems from our Native ancestors being forcefully removed from the land, slaves stolen from other continents, refugees from areas that the US destabilized or underdeveloped, and immigrants. Here, we are united under one flag because we believe in opportunity in the New American Dream for everyone.

Being Mexican American means I am born or raised in the United States and living with American pride in my mind while I have Mexican blood flowing through my heart like the Rio Grande. Mexican Americans love corridos and cumbias, yet also love listening to Kendrick and Tupac and eating hamburgers while setting off fireworks on the Fourth of July. Our Spanish is imperfect, and our accent is evident when we speak English; better yet, we have merged them to create the famous Spanglish. Our Mexican family members frequently criticize our Mexican family members for being too Americanized while also being put down by our American peers for being too Mexican. As Americans who have an immigrant background, we struggle in this weird in-between space since we are involved in a constant identity conflict. We have to live and exist among two different worlds, the old and the new because we are from here and there simultaneously.

Going from Colonization to Alienation— *and Why It Must Stop*

★ ★ ★

After the colonization of the American continent, the settlers used an additional tactic to alienate the people who originally belonged there. The most obvious example today is when I get told to "go back to my country," or when the love we have for our culture is called "foreign" by others. When people say this, they are ignorant that this was our land, this is still our land, and we shouldn't go anywhere because we are Native Americans living on colonized soil. We can't go back to our country because we are already in our country. Mesoamerica is our ancestral land, and as such, we have sovereignty over our inmigration rights.

Instead of accepting other cultures, old America attempts to forcefully assimilate groups to destroy their cultures in service to white supremacy. Culture is an essential aspect of the human experience. Power and privilege determine how long the culture will live healthily and survive. This section's primary objective is to describe the plight of Native Americans through my perspective and how I have struggled against power inequities in the context of postcolonial alienation.

As we have learned, by virtue of power inequalities and power abuses in society, the people in power can determine what is authentic, accurate, and deserving according to their standards. As a result, they can easily neglect to acknowledge what is important to other groups of people. When this happens, the people without power will receive differential treatment from systems and institutions created and designed to concentrate power on one group and neglect the rest. Through socialization of the oppressed group, there also tends to be an internalization of inferiority, which manifests itself at the biological and psychological level. In other words, the colonization of the oppressed group occurs through several decades of socialization, and the internalization of this socialization has happened throughout history.

As First-Generation Americans (FGAs), we often do not learn the history of our ancestors in traditional education. We understand that civilizations were decimated after colonial conquest, that the Native Americans were all wiped out. But I'm afraid I have to disagree. The descendants of Native peoples in America are still here, more resilient than ever.

Before the bloody and violent conquest of the New World began, at least sixty-two Indigenous groups and languages spanned over three thousand years of history from 1500 BC to 1519 AD in America, with the biggest tribes being Aztecs and Maya.[1,4] They had a complex and highly developed written and spoken language, K'iche', and had territories in Guatemala, Mexico, Honduras, and Belize. During the 15th century, the conquest was unmerciful, and much knowledge generated by our ancestors was destroyed or written out of history. This goes back to the fact that those in power determine what is standard and accurate—the victors write history. And this perpetuates a vicious cycle because knowledge is power. By eliminating this ancestral knowledge, the current power structures have stripped these communities of their power.

The beginning of the Native American Holocaust began in 1492, and the conquest of Middle and South America began in Veracruz, Mexico, in the 1510s.[1] I will not say the colonizers' names who attempted to exterminate the Natives and enslave the Blacks. Many argue that they were successful, but others say that they were not because we are still here. In the north, several Englishmen massacred and forcefully resettled several Native tribes, including the Apache, Iroquois, and Cherokee. Despite several hundred years of explicit discrimination and power exploitation, the Cherokee could not become recognized as US citizens until the early 1800s. In other words, it took over three hundred years for them to become citizens in the land that once belonged to them. However, this isn't just unique to the Cherokee. There are still Native American people who are struggling to gain citizenship, in particular, Native Americans from Middle and South America.

In Middle and South America, Native peoples faced similar situations as their neighbors in the north. After the Spaniards betrayed the Aztec empire, colonial forces expanded south toward the Maya region and attempted to conquer their people and steal more power. Many Mayan people fled to the mountains in the rainforest for resistance and survival. They ended up resisting for two hundred years before the Spaniards declared them under their control. That is why, today, their culture has persevered, and it remains alive and cohesive, unlike many other Indigenous cultures in Mesoamerica. Or at least that is what we are taught to believe. I repeat we are still here.

The Native Americans' immune defenses succumbed to the epidemics that came from their colonizers. During the early colonial period, smallpox, measles, and mumps took a substantial toll on Native health during the conquest. During the entire subjugation of America, about twenty-four million Indigenous people died in only the first sixty years of the first encounter – which accounted for at least 10% of the worlds population.[7] The mortality and morbidity rates associated with America's conquest continue to be substantial to the point where some communities were utterly exterminated, and some are still experiencing the consequences today. In comparison, forty million people died during the Black Plague by 1353, and eleven million were killed during the Holocaust in 1945.[8] Manifest destiny was the main reason early American founders justified stealing land and their attempts to exterminate others to fulfill what they believed was right. The first colonizers brought disease to the Americas; they also brought wrath, weapons, and ideas that continue to harm our fellow human beings.

History has taught us that the colonial imperial forces brought trauma to several societal segments by abusing their power and privilege. Historical trauma is the cumulative emotional and psychological wounding over the lifespan across generations. The increasing exposure of over five hundred years of repeated trauma through structural and oppressive systems of power continues to cause physical, emotional, and social harm to people today who have to live through various biopsychosocial mechanisms that are still not fully understood. Although we are only starting to understand the root causes of structural pathologies, we know that the damage has already been done and that there has been a substantial impact on people's health and well-being. Several hundred years after the conquest began, power began to give birth to racism and started to make us sick. In our country, it is evident that our immigration system was built on racist ideologies, which have contributed to various racial health inequities. By ignoring what history tells us, today's racial health disparities continue to be a form of genocide of non-white people across America. Racism, not race, produces our most intractable health inequities, and historically, immigration policy is one of the main ways we uphold and sustain these inequities.

Why Immigration Makes America Great

★ ★ ★

Immigration has been happening since the origins of human life. The US is often portrayed as a nation of immigrants and is known worldwide as the country that fights for freedom and the opportunity for everyone to pursue life, liberty, and justice. However, currently, this notion can be nothing further from the truth at this moment in time. Many people fail to realize that life, liberty, and justice mean equitable education opportunities, security and safety, jobs, and health care. In the US, there is strong opposition against immigration, but why? The short answer is not about resources, since we are the wealthiest country on earth, but it is about protecting the ideology and foundations of white supremacy in the US. Many people in the US and other parts of the world do not understand that immigration is something that cannot be stopped and is something that should continue to be embraced, for the reason that diversity is what makes a country great.

It is problematic when we consider that today, in America, we live in institutions built by African slaves and on Native land at the expense of lives primarily for the benefit of white people. In turn, who has created our immigration system not to enhance diversity but rather to attempt to maintain a predominantly white and wealthy population in the US. Rather than fighting the corporations who steal millions of dollars, Republican lawmakers tell us that the heart of our country's problem is immigrants and crime. Some argue that immigrants waste resources and take people's jobs, although the reality is immigrants agree to work at meager wages with minimal to no benefits. Instead of attacking the person who is working for a low wage, why don't we ask ourselves why the people who are hiring them are paying such low wages in the first place? Immigrants make America great because they are the cornerstone of our country's success. Anti-immigrant rhetoric is rooted in America's deep and dark history of racism. Immigrants will have to learn that racism is nothing but an antiquated ideology with no scientific, moral, or ethical grounds for the inequities it produces.

As we tackle immigration head-on in the coming years, we must ask ourselves as FGAs how we can build the best environments for the next generation. As we have learned, generational trauma is real and has significant consequences on many people's health, especially those who have had power stripped away from them. Therefore, we must first understand how our immigration policies impact health. We must also commit ourselves to see and embracing each other's color and appreciating all humans' inherent diversity.

When we talk about the next steps in dealing with immigration, it is essential to address white privilege pertaining to immigration. In particular, we must be critical that those who have colonized most of the world also have the highest degree of freedom and mobility to migrate anywhere they would like. At the same time, the rest of us face the strictest restrictions to visit this country. This is the form of white privilege in immigration. Some people become offended when we refer to white privilege because they believe it doesn't exist. The truth is we are not asking anyone to feel guilty for the atrocious actions their ancestors have imposed on our ancestors. However, we are requesting mutual recognition that the system was built in such a way to benefit mostly white people at the expense of others, and we are demanding follow-through and next steps to the right for this wrong. We are requesting that we work together to dismantle these systems of oppression, not to rebuild them to favor any particular group but instead to focus on equity and restorative justice. Immigrants from all walks of lives must rise from the margins and demand the promise that America claims they give to all of us.

Throughout my life, my medical education, and my advocacy, I have consistently realized that my mission revolves around learning my roots to ensure that I can provide the highest quality of care to my community. I have realized that the social determinants of health in immigration are root factors in some of our most intractable health inequities. Still, we cannot address such systemic factors independently. Instead, we must work together across multiple sectors and collaborate with others from different fields. For example, in my case, I have been working alongside my patients and leading officials to combat health issues by creating novel interventions and prevention strategies at various levels that can move us toward the elimination of racial health disparities. I have been actively involved in my patients' care and legislative sessions by testifying and informing decisions that impact population health. Demanding basic life needs such as job security, access to high-quality and free education, and having safe neighborhoods. I do all this to ensure that the next generation can recover from the historical events across the American continent.

Now that we are armed with our history and have a better idea of how we got to where we are, we can begin to discuss what we can do next and dissect and repair the injuries that the people in this country have endured. As the incoming generation of Natives, we –don't only administer health care but also mentor. We have been called to fight oppressive structures in our society and do the diversity work when our allies' emotional effort has been drained. As we begin this new era of addressing systemic issues, we must focus on education, housing, hiring, and immigration to weed out sexual harassment, colorism, ableism, and several other isms. To start, we must oppose any anti-immigrant rhetoric rooted in racism.

REFERENCES

1. Stannard DE. *American holocaust: the conquest of the New World*. New York: Oxford University Press; 1993.
2. Ancient Mesoamerica. 1990 (Journal, Electronic).
3. History. Guerrero History. *History Topics Mexico* 2018;1.
4. Coe MD, Urcid J, Koontz R. *Mexico: from the Olmecs to the Aztecs*. Eighth, revis and expand ed. London: Thames & Hudson; 2019.
5. Todorov T. *The conquest of America: the question of the other*. 1st ed. New York: Harper & Row; 1984.
6. Navarro A. *Mexicano political experience in occupied Aztlan: struggles and change*. Walnut Creek, CA: Altamira Press; 2005.
7. European colonisation of the Americas killed 10 percent of world population and caused global cooling. *Space daily*. 2019.
8. Newson L. *Devastation: The World's Worst Natural Disasters*. 1st American ed. New York, N.Y: DK Pub; 1998.

PART 2

Translating Theory into Practice through Storytelling

The Life Course

Using life course theory, adverse childhood experiences, and biology to understand the social determinants of health

> *"Only by digging deep down to the core of our true self can we come to a place of inner certainty. Our underlying values and priorities are our personal navigational starts on the life's journey—essential tools to chart a life course that embraces what matters most to us."*
>
> —*Marian Deegan,* Relevance: Matter More

The Life of Remedios Montiel Garcia

★ ★ ★

Remedios was born on September 1, 1954, in Acapulco, Mexico, where she spent a large portion of her life. She was raised in a poor, traditional Catholic Mexican household on Acapulco's outskirts, which are still known as colonias or colonies. Mexico's hierarchy considered Remedios's family as lower class in Mexico's social and economic order because they were of mixed heritage, mainly Spanish and Native American. She started hitting roadblocks at a young age. She could only make it through six years of school and dropped out at twelve years old to help support the family by picking up a full-time job cleaning houses and hotel rooms. Money was a top priority, but soon her priority became her family. She and her first partner were never married but lived together for seven years. He was from a small town in Mexico but moved to Acapulco for the tourist industry to work at the hotels. When she first started dating him, Remedios was sixteen years old, and he was forty-six years old. For the next several years, they worked together and even had one daughter. At seventeen, Remedios gave birth to her first daughter, Monica. However, considering the vast age differences, she experienced a lot of trauma from him, and it was also easy for him to be promiscuous. She felt like she had been taken advantage of and knew that he was not the right person for her.

Several months later, she decided that it would be a good idea to keep working and find love elsewhere. In Mexico, it was a common tradition and expected that you work to make ends meet and have a family, and it is sometimes still the case today. Shortly after, she met a man whom she fell in love with. Unfortunately, when she became pregnant, he left her. There wasn't a single moment that Remedios stopped working. All she could think of was how she was going to take care of her first daughter as well as her new baby on the way. At the age of twenty, she had her son, Rogelio. It wasn't easy for her to raise two children on her own, both in terms of time and finances. She had to fight against poverty, so her kids could have a better life in the future since they were living in deprivation.

She first married at thirty years old, and this was her longest relationship, lasting twenty years before he passed away in 1997. They worked hard together to sustain the family. At the age of thirty-two, she had another daughter and her last child, Monserrat. Remedios was always dedicated to ensuring that her children were taken care of despite their poverty. However, all three of her kids also fell into the cycle of poverty and marginalization and couldn't complete high school. Everyone in the household had to work a job to pay the bills and put food on the table. Her children would work full-time jobs, sometimes two jobs at once, so they could support the family because that was all they ever knew to do. After they started working and making enough money to support each other, they began to stabilize their lives and plan for a brighter future with the hopes of someday returning to school to get an education.

In Mexico, Remedios was healthier because she was also younger, but at the same time, she would not take care of her physical health throughout her life course since she did not have the time to exercise and be active. She ate a diet that only consisted of processed foods and beans and rice. It wasn't until she was much older than her health started to decline rapidly at the age of forty-five when she came to the US.

Remedios's life forever changed when Monica immigrated to the US. Her first child was the first in the entire family's history to become a US citizen named William Mundo. Remedios ended up following Monica and immigrated to the US and settled in Los Angeles to be with my mother and me. When she was living in California, everything started to piece itself together, and she began to make the best income she had ever made in her life. Eventually, several people in my family lost their jobs due to the economic recession in 2000. We started to struggle again for a few years. Remedios and Monica worked three jobs to make a decent income to support the family here and Mexico. We lived in a small one-bedroom apartment that housed nine people. It was crowded, but we all had to live together to have enough to pay for rent. After years of living in California and supporting the family, employment in the area began to decline. The family had to move out of the city to a rural mountain town in Colorado to work in the ski industry, clean houses, and hotels. After many years of experience, Remedios finally moved up to the supervisor level, where she got to do less labor-intensive work. However, many days, she still worked harder than the housekeepers and was still able to clean efficiently.

In my grandmother's case, we can see the fundamental causes of disease and the preexisting conditions in her physical and social environment. She could not access any care during her life because all she knew to do was work for survival. In her entire life, there were more times when she was under allostasis instead of in homeostasis. She had several diseases in her life, including high blood pressure, high blood sugar, and rheumatoid arthritis. Her lack of access and necessary financial and social resources were other fundamental causes of her diseases.

Due to economic inequality, she always had to work and could not go to school to pursue her own goals. It is well known that education is directly correlated to socioeconomic status, which influences health status. Being uneducated and raised in a low-income family led her to engage in behaviors that affected the health outcomes she later developed, such as working long hours in labor-intensive jobs, a lack of nutrition, and constant stress. Due to social disadvantages in Mexican culture, my family was marginalized, and resources were denied merely because of our heritage. Some of the resources taken away from us included access to education, access to support services, and access to opportunity. Several of these resources were denied by the people in power—in my grandmother's case, it was white Spanish settlers. One can imagine that living a life where they are continually fighting against institutionalized discrimination can take a toll on their health, just as we have learned with allostasis.

The primary way my grandmother experienced allostasis through her life course was through role overload, which we learned is when the demands of the role placed on an individual exceed that individual's ability to meet them. There were also many inter-role conflicts, which occurred when the requirements of two or more roles held by her were incompatible, such as having to raise two children on her own as a young working mother. It was difficult for her to meet her demands when she had to play both the role of a father and mother in raising her children. The stress from these demands follows the stress pathway's direct route when chronic strains influence disease outcomes.

Since my grandmother was a geriatric, low-class, Mexican woman in the US, in addition to her background in Mexico, she was likely to have higher stress-related mental and physical health problems in her later life. But although she had several health challenges in life, there was never a time when she felt withdrawn or embittered. Instead, she was always inspired to do better and allow her children to do better. At the end of the life course interview, she said that everything she had been through was worth it. My mother, uncle, and aunt all agree that they have had better opportunities in life because of my grandmother. Because of my grandmother, I have also been granted the opportunity to better my life and better my children's lives.

My interview with my grandmother highlighted how the life course could be an essential way to understand better the lives of the people we love. The life course is also a useful tool to discover how structural pathologies manifest in our lives across our lifespan. I can still remember that tears started pouring down her cheeks toward the end of the interview, and she hugged me, saying that she couldn't wait to see me begin medical school before she passed away. Watching me become a medical doctor was something she always dreamt of every night. Although my grandma was a relatively young senior citizen, her health did not reflect her age because of the historical and structural pathologies that have resulted from her life.

The Life Course

★ ★ ★

The life course approach, also known as the life course perspective or life course theory, refers to a method produced in the 1960s for analyzing people's lives within structural, social, and cultural contexts.[1] A life course is an approach to health that addresses the social determinants of health (SDOH) during specific developmental periods of life. The life course can illustrate the different risks or protective factors a person experiences and how they can have a cumulative effect on a person's health over their lifespan. Hence, it is vital to use the life course when studying the SDOH and the collective impact of inequality throughout the lifespan to improve people's health. The life course is measured through various individual life experiences by collecting fundamental health indicators and stories from the person on how they take care of themselves and any challenges they have faced in the past. Different indicators include low birth weight, stunting, reduced physical labor capacity, lower educational attainment, inadequate food and health care, shortened life expectancy, and restricted economic potential. All of which are fundamental causes of disease.

The life course includes pregnancy, early childhood, adolescence, adulthood, and older adulthood, all of which are different moments in which people's health status can change. From the emerging field of epigenetics to adverse childhood experiences, we are starting to learn about maltreatment mechanisms and later-life health and well-being.[2] We have learned that people living through challenging conditions in the US are at significant risk for the leading causes of illness, death, and poor quality of life. Some of the types of traumas that people can experience during different parts of their life include verbal, physical, or sexual abuse and family dysfunction, and these traumas have all been correlated with substance use, depression, cardiovascular disease, diabetes, cancer, and premature mortality. In the life course and epigenetics field, several studies have demonstrated that long-term adversity contributes to the onset and progression of the disease.

Using the life course, we will see that many of the nation's worst health and social problems arise due to historical trauma, epigenetic regulation events, and adverse childhood experiences. The life course allows us to chart individuals' progress and identify the places where we can start to prevent and treat these problems.

Homeostasis and Allostasis

★ ★ ★

Stressors and events are circumstances that cause unease or physiological responses to competitions. The process in which the mind-body attempts to deal with discomforts to recapture a sense of homeostasis is called allostasis. In biology, homeostasis is a state where the body is in balance in terms of physical and chemical conditions maintained regularly by your organs. Humans strive to be in homeostasis because it serves as a baseline for everyday health. Some of the different factors with which we can measure if someone is in homeostasis include body temperature, heart rate, respiratory rate, and the specific amounts of nutrients and fluids in your body. Here's an illustration of how humans achieve homeostasis: you are in the middle of the desert, burning hot. Your body will sense an increase in outside temperature. So to maintain homeostasis, your body will start to sweat to let go of all the heat in an attempt to return to a normal baseline temperature. Now, imagine that your body cannot reach homeostasis and unable to go back to a baseline state—this is allostasis. In other words, it is the opposite of homeostasis. Often, the primary way that allostasis is maintained is through constant exposure to stress. Stress is a common cause of emotional body disturbances and psychosomatic illnesses, and it can even worsen other diseases.

In our body, the stress pathway that begins allostasis starts in our cerebral cortex. Our cerebral cortex is the most outer part of our brain, and it is responsible for higher-order functions. After our cerebral cortex becomes activated from the stressor, our limbic system, which is another specific part of the brain that controls emotions and memories, is activated. The limbic system then activates our hypo-pituitary-adrenal axis. This complex hormonal response activates our adrenal glands that sit on top of our kidneys to secrete hormones that trigger a bodily stress response, causing your nervous system to arouse your body. The primary stress hormone often implicated during the stress response is cortisol, which is released from your adrenal glands. Cortisol has many effects on the body, mostly good because it is vital for our survival. However, when we have too much cortisol, we begin to see problems in the body's ability to maintain homeostasis. We know that the body likes to exist in harmony between the mind and body, usually in a calm state where cortisol is not excreted in excess. But when the body feels like it is continuously under attack, such as in cases of chronic discrimination, our nervous system begins to overwork. As a result, there is an increase in cortisol, which leads to the body doing different things, such as increasing our heart rate and blood pressure, which can be harmful at chronic levels. When a stress response happens in a short-term setting, such as when you confront a bear, your body goes into the "fight or flight" response. This automatic response is essential for our survival. However, in some cases, especially for people who experience discrimination daily throughout their lives, they are frequently activated, and overtime, this causes negative impacts on the body. As a result of allostasis, or in simpler terms, the chronic stress response, the body begins to adapt to this constant heightened state, which in the end causes exhaustion and disease.

When the human body is experiencing allostasis from social stressors, the disease tends to strike during these vulnerable moments. One way allostasis facilitates the development of the disease is through immune dysregulation, which also causes a proinflammatory cytokine response—the chemical signals in your body that cause inflammation and damage. During immune dysregulation or when your body is under stress, such as when you are sick, your immune defenses are not as robust. In many of these cases, we know that immune dysregulation enhances infection risk, prolongs infections, or even delays wound healing.

Now, going from a microscopic, biological level to a broader perspective that includes psychology and sociology, we can see how allostasis can play out in our daily lives. The life course is especially useful in this area because we can conceptualize how life stressors are correlated with the long-term problems, conflict, and threats that we face regularly. Some of these social stressors include the different roles people find themselves in and the issues they can cause. For example, when overwhelming demands are placed on an individual who cannot meet them all, when you are in an unwanted role or when you experience changes in long-term rules, conflict with others, or even when you have two conflicting desires. These are all examples of social stressors and ways in which stress can impact our lives.

Once you understand how emotional stressors can physically affect the body, you can begin to understand the harmful role that racism, sexism, classism, etc. can inflict on the body from the chronic stress response that our body has to maintain. Through using life course theory, we can understand how people's daily interactions can influence their health outcomes in complex ways over time. We can also use it to document how generational traumas impact health, as discussed in part one with epigenetics. By examining people's life course, we will be able to chart health outcomes through the generations to come. I encourage you to learn more about life course theory and how you can incorporate it into your daily practice. It is easy—all you need is to find the time to listen to someone else's story across their life by conducting an interview. You can prepare some questions ahead of time and even share with the person you will be interviewing. I recommend asking for permission to record the interaction so you can listen to it later, documenting it, and then even conducting a life course analysis. In the next part of this chapter, I walked you through a life course analysis, I was able to do when I was a college student.

In this chapter, we have learned how we can utilize different areas of public health sciences to illustrate people's life journey and the significant events that occur throughout life that can have an impact on health. We understand the biological and psychosocial pathways, and we have discovered how to translate the topics we studied in previous chapters into concrete examples. The life course of my grandmother is just one example of these theories translated into practice. Moving forward, the rest of part two of the book will highlight other concrete examples of the concepts we've learned through the telling of my own life and story.

REFERENCES

1. Cheng TL, Solomon BS. Translating Life Course Theory to Clinical Practice to Address Health Disparities. Maternal and child health journal. 2013;18(2):389-395.
2. Merrick MT, Ford DC, Ports KA, Guinn AS. Prevalence of Adverse Childhood Experiences From the 2011-2014 Behavioral Risk Factor Surveillance System in 23 States. JAMA Pediatr. 2018;172(11):1038-1044.

The First-Generation Mexican American Physician

My journey from experiencing health inequity to my calling in life

New-American Healer

★ ★ ★

As a teenager, I woke up at 5 a.m. during the weekends to catch a forty-five-mile bus ride over the Battle Mountain Summit pass in Colorado to go to work with my grandmother. My grandmother worked two jobs as housekeepers for ski resorts and cleaned houses for private clients. There was never a time when my family was not working, and they insisted that I learn how to start to work as well so they could better prepare me for life. As I stood anxiously at the bus stop in -20°C temperature, awaiting the reassuring sound of the approaching bus, I remember how the bitter-cold air and severe wind during the winter would penetrate my bones and freeze my eyelashes every time I blinked. I understood that I needed to do this because my family had emphasized that we must never take for granted any opportunities that we had, since we hadn't had many.

It was not my favorite time of the week; however, my grandmother kept me inspired by reminding me of our family's immigration journey in the pursuit of the American Dream. I still remember how my grandma would stand there with me, telling me how we must work hard and commit our allegiance to this land. "After all," she said, "this is your ancestral home." She would tell me that I should never be ashamed of my background or my accent when I speak. It was insisted that I learn English, but they never discouraged me from speaking Spanish. It was how they preferred to communicate since their English wasn't great—and neither was mine. She would say to me to always appreciate the two countries that made me who I was. She would tell me this as her warm hands rubbed my cold, pink cheeks and red nose in the freezing weather.

My family left Mexico in the 1990s to pursue the American Dream and to give our family life-changing opportunities. Several of these opportunities have evolved into my pursuit of a career in medicine. Since I was the first to be born in this country, I was determined to take care of my family. Since we lived in poverty, several people in my community became sick frequently. When people got sick, it didn't mean anything to me. It wasn't until my own family members started getting sick when I became not only concerned but also curious.

Whenever my family had an encounter with the health-care system, I had to interpret from English to Spanish for them since they were unable to communicate with their providers. With no access to health care, my grandmother suffered many episodes of high-blood sugars and high-blood pressures. These episodes happened daily and sometimes every other week. This was especially evident when we were unable to afford her medications and visits with her providers. I was eager to begin my career in medicine to finally provide for my grandmother and reciprocate the compassion and love she had raised me with. As a young child, my grandmother and my family had always relied on me to be the primary contact, in particular when it came down to vital health, legal, and financial decisions. I was determined to be prepared and to learn how to continue caring for my family.

While I was in college, my grandmother suffered a premature death at the age of sixty-two from a ruptured cerebral aneurism and a subarachnoid hemorrhage. A couple of days before, she had been in a car accident and hit her head. However, when she went to the hospital, no imaging or further work-up was done on her head injury since it was dismissed as a mild concussion. I wonder if they had scanned her head that day if they would have been able to prevent what happened.

When I walked into the hospital that day, my mother and uncle frantically shook me to help them understand what had happened. I was unsure what was going on. I was in school but I wasn't even a doctor, though they thought I was. I felt lost—the only thing I knew was that she had been on Flight for Life and was on a mechanical ventilator. Flight for Life is when a helicopter needs to fly you to another health facility that is able to save your life. Instead of going by ambulance, she had to be flown down to Denver from the mountains to be taken care of. Her condition was critical, and after two brain surgeries, she was finally left in the medical intensive care unit, where she was completely unconscious and only responsive to pain.

I still remember the sounds of the machine blurring with my family's cries as I looked at my grandmother's swollen face. I was forever changed after that day. It was then that I felt like I had let my family down since I had been unable to adequately answer their questions and be the person they had always counted on me to be.

Although I was only a pre-med student, I was already a healer in the eyes of my family and community.

Feeling helpless, the only thing I thought to do was to speak to the doctor to understand what was going on. As soon as I found him, I asked if he could please explain why this had happened to my grandmother. The doctor came closer to me with his arms folded. He tersely replied, without any additional detail or empathy, "It was just random chance." The physician did not sit down with us to discuss why this might have happened, nor did he offer to speak to my family with an interpreter. The first time my family received any type of information was from me,. Embarrassed, I said, "They don't know what is going on; it was just bad luck." After our short encounter, I remember watching the physician go to the room across from my grandmother's, and his tone and compassion immediately changed. I couldn't help but wonder if my family had been treated this way because we were Mexican immigrants. I felt so disappointed by the medical community because I felt like my family had been harmed by the people I aspired to be like in my pursuit of the New American Dream.

After several years of not fully understanding the pathophysiological and public health forces that caused my grandmother's demise, finally, I remember sitting in my neurology school class anxiously waiting to learn about the causes and management of subarachnoid hemorrhage. Although her chances of surviving were not high, I learned that, if managed well, people can recover, though not entirely, which I understood. However, I had been under the impression that no one survived from a subarachnoid hemorrhage because the physician had insisted that my grandmother was dead and that there was nothing they could do for her anymore. But we did not know why. He had had no empathy for us.

That day, when talking about our next steps and what we wanted to do, the physician asked my family how we were going to pay for her care. He asked about our insurance status but quickly changed the topic when he heard us speak Spanish as we tried to discuss what we were going to do in order to afford all of the care she had received. "If you don't have insurance, it will be impossible for us to pay for the advanced medical care she would need," the doctor said. We contemplated so hard. We were unsure of what to do. A couple days after, still with no recovery, we met with palliative care, which is the team that takes care of patients who have life-ending conditions. At the end of the discussion, one of our options was that we keep monitoring her to see if she would improve, knowing that even if she did improve, she would have irreversible brain damage that wouldn't let her live a normal life. The other option was to move forward with unplugging her from life support, which was the only thing keeping her alive. My mom felt overwhelmed. She looked at me and said, "You decide what we do; you know best."

It took me a while to think of what to say or do. I had never in my life had to make a decision of such proportions. After reflecting on my grandmother's life, I knew that she would have preferred not to be on life support. Every day, she would be suffering from her medical illnesses. They were destroying her alive. I was confident that my grandmother would have wanted this, and that was the most important part of the decision that I made that day.

Sometimes, now, I wonder if she might have been alive today if I had known more about her condition and how to best advocate for her at the time. Would she still be alive if her clinicians had not known that she was an undocumented immigrant or that she was uninsured? Would she have received better treatment if my family had been able to communicate with her health providers? As a medical student now, and when I reflect on how the physician treated my family, I swear that I will never let that happen in my medical practice and that I will contribute to society by becoming the physician that America needs.

Increasing Diversity in Health-Care Workforce

★ ★ ★

By increasing diversity in the health-care workforce through increasing the number of diverse health-care providers and programs, we can begin to eliminate health disparities and help serve historically and medically underserved patients with the respect and dignity they deserve. Not only because they are completely worthy of it, but also because they have helped build our country. On top of caring for others effectively, we can also increase the opportunity to have more First-Generation students pursue a career in medicine. Over and over again, we have seen how First-Generation medical professionals have the potential to make a significant impact on our world and those around us. First-Generation medical professionals can provide unique benefits in the health-care workforce because we have the ability to be culturally responsive, as we have discussed in previous chapters. However, it is important to recognize that even some FGAs can come across as arrogant, such as the doctor handling my grandmother's care, and not all non-FGA doctors are arrogant. It is important to have diversity in the health-care workforce just as it is equally as important to have better training in cultural sensitivity for all health-care workers.

One clear example in which I have learned to practice culturally responsive care was when I was working with my preceptor at Denver county's safety-net hospital. I was asked to take care of a Spanish-speaking, sixty-four-year-old Latina woman who had been diagnosed with stage IV marginal B-cell Lymphoma. This is an aggressive late-stage tumor that is deadly; however, if the tumor is caught early, it can be curable. I was tasked to discuss how she was feeling, treatment options, and hospice care. As I nervously stood outside the exam room, preparing my notes and questions, I noticed that the medical staff was having a hard time engaging her and her daughter. When they asked questions, her daughter interpreted for her mother, giving me flashbacks to the times when I would do that for my family. Unfortunately, the daughter was unsure of how to tell her mother that she had less than six months to live.

As soon as I walked in, I quickly introduced myself in Spanish and sat next to the mother and her daughter. I told them that I was part of the team that would be taking care of them that day. After we talked about her life back in Mexico, her favorite grandchild, and her current condition, I couldn't help but share my personal experience with death in my family. She reminded me a lot of my grandmother, and I shared that I wished I could have been able to have had this talk with my grandmother and her provider. After careful consideration of her goals, she said that she would be most happy with hospice care and getting off chemo. She could happily spend the rest of her time with her family without dealing with all the adverse side effects of treatment. Before I left, I offered my warmest regards and wishes and said that I would be right here if they needed anything. As I walked out, her daughter quickly got up and followed me outside the exam room, screaming, "Muchas gracias, Dr. Mundo!" I proudly smiled and told her she was welcome but that I wasn't a doctor yet. She quickly replied, "Soon," with a tear coming down her face.

It was a great feeling to be able to form a therapeutic alliance with my patient built on cultural responsiveness and linguistic congruency. She was discharged later that day, happy that she could go home and spend time with her grandchildren. One month later, her daughter notified us that her mother had peacefully passed away.

As I prepare to finish medical school, I realize that this is the moment I have been waiting for since my grandmother passed away. The opportunity to walk down the halls of the hospital, carrying the accents of my last name on the white coat that I wear with grace, and providing comfort and empathy to families like mine, making America great. My goal of becoming a doctor started long before I had family members who lost their lives to medical injustices. These experiences have only solidified my goal and also allowed me to reflect on when my journey to medicine even first started.

Inspiration from Immigration

★ ★ ★

My curiosity in medicine was first sparked when a family came knocking desperately at my father's trailer doorstep. The first thing I noticed was a couple carrying some firewood. Behind them, I could see a timid boy with his hands wrapped around his belly, a sign of pain. Whether for money, firewood, or a smile, my father was always more than happy to help. I grew up watching my dad provide care to our community, and not just care in the sense of seeing people when they had medical concerns; he was also there to fix things around the community and provide shelter and food to others. He always had a hand out to help others and was well respected.

When the family came inside the trailer, I decided to peek through the door, where I saw my father care for them as if they were his patients. It reminds me a lot of the primary health-care model where family doctors are placed in the communities in which they serve. It was at this time that I believed he was a physician. It was not until I was much older that I discovered my father was not in fact a physician; he had not even passed the sixth grade. Even without a formal education, he used his knowledge of indigenous practices and served as a Curandero to heal our community. A Curandero is a type of community healer in traditional Mexican culture who specializes in the alignments between physical, emotional, and spiritual balance.

Before becoming a healer, he left everything behind in Mexico and immigrated to the US to provide me with a better life. After watching his practice from a young age, I knew I also wanted to provide care to the people around me. When I saw him be there for others, I saw how it was a privilege to be a person others look up to and whose care they can ask for when they are in need. These are my reasons why I aim to impact my community, country, and the world positively. I want to be a healer like my father and provide relief to those who suffer. As I reflect upon my dad's sacrifices to ensure that our family's and community's needs were met, it solidifies my drive to become a public health–trained physician myself. My passion for medicine stems from a deep commitment to health equity and social justice, as I mentioned in part one. I believe that an individual's most valuable possession is their health. With that in mind, all individuals must receive a means to seek medical attention. I believe that healthcare is a human right as good health facilitates a long life, liberty, and the pursuit of happiness.

Although I always struggled with the sciences in high school, it wasn't until I got to college that I started to excel in my science courses. I was completely blown away and fascinated by what the human body does. I started my undergraduate career with a major in biology; however, I later decided that I could major in whatever I wanted if I planned to go to medical school. All I had to do was ensure that I finished my pre-med requirements. After learning more about my history and identity, I was introduced to public health sciences around the same time. I still remember sitting in a classroom when everything clicked, and I rushed to set up an appointment with my advisor to change my major to something that I had a stronger interest and passion in: ethnic studies and public health. While I did not major in biology, I still had a strong record in the sciences since I took a good foundation of science courses to prepare me for medical school.

I decided to pursue a career in medicine because I wholeheartedly enjoy the fruits of scientific inquiry and cutting-edge biomedical technology. However, I also equally valued the opportunity to stand in pragmatic solidarity with all people regardless of race, religion, color, creed, or sexual orientation. With the demographics of our country changing rapidly, there is a need for bilingual and culturally responsive health-care providers. Our heritage will help address the rising demand. I have learned that it is important to strive for diversity, inclusion, societal change, and social responsibility in a multicultural society.

Through my undergraduate career, graduate career, and professional career, I channel my First-Generation and New American mindset to set myself up for success in the field of medicine, public health sciences and critical race/ethnic studies. As a First-Generation student, I invested my time, knowledge, and skills toward these movements and causes because they equip us with the unique ability to connect with people. Through these experiences, I have been able to not only obtain the education I have earned but also create action and change.

I have observed that in America, it takes a lot more than passion to combat the root causes of social ills. As First-Generation medical students and professionals, we must emphasize and treat health equity as a foundational science to address the social determinants of health; we must conduct scientific research to identify evidence-based interventions that work for everyone and not just a few. All the things we have learned will allow us to advocate for universal health-care access, basic needs such as job security, access to affordable and high-quality education, and access to safe neighborhoods. In this chapter, I have demonstrated the impact that First-Generation physicians can make in our communities. We can be those who go beyond treating patients to addressing systemic and macro-level pathologies that impact patient populations by using policy, advocacy, and development.

As First-Generation medical students and professionals, we must be motivated not just to serve community health-care needs and impact other people's lives but also to discover different and holistic approaches to the way we look at health. With the globalization and urbanization of our world, as well as the transition from infectious to chronic diseases, we must continue to meet the needs of the evolving communities we serve. So far, we have learned that we can accomplish this by broadening the focus of medicine from treating illness to one that equally emphasizes prevention. We will have to become active and invested participants in the lives of our patients and communities. Additionally, we must encourage and challenge community members to develop leadership skills and to understand the significant intersections between public health and medicine in their own lives as well.

The War on Drugs and the School-to-Prison Pipeline

A case study of my personal experience with the criminal justice and school system

> *"The nature of the criminal justice system has changed. It is no longer primarily concerned with the prevention and punishment of crime, but rather with the management and control of the disposed."*
>
> —Michelle Alexander, The New Jim Crow:
> Mass Incarceration in the Age of Colorblindness

My Experience with the School-to-Prison Pipeline

★ ★ ★

When I was sixteen years old, and I remember a cold autumn night in the middle of October 2014, the night before one of my high school football games. This was the night I saw the school-to-prison pipeline unfolding in front of me when it felt like my life was ruined.

I still remember my mind running in all different directions, my heart feeling like it was about to beat out of my chest, and as if a baseball were stuck in my throat. However, the worse part of all was hearing my mother cry. As soon as I opened my eyes, I could see the flashing red-and-blue lights outside my window and two large men with their hands on their holsters. I was confused and disoriented, and I had no idea what was going on. I thought I was just in the middle of a bad dream for several minutes, like a lucid dream, you can control and where you have autonomy over your unconscious thoughts and decisions. Reality didn't hit me for several minutes; it wasn't real until the cold air from the snow outside the trailer rushed in and crept up my legs. I was standing in my boxers in the cold, humiliated, and trying to figure out what was going on. One of the officers screamed at me, asking, "Where are all the drugs?" I could hear my mom crying in the background, screaming at me, asking me what was going on. I didn't know the answer; all I knew was that I was traumatized.

I was so confused. For several minutes, I insisted that they had the wrong person. It wasn't until the police officers started asking several questions about my friends and my activities when things began to make a little more sense. I wasn't sure how they had even gotten into the trailer to start. It turns out that when they had come to knock at my door at two in the morning asking for me, my mother was concerned and wanted to make sure I was safe, so she let them come inside. She couldn't understand what they were saying, but it appeared that they were making it seem like something had happened to me, so my mother immediately welcomed them into the house and took them to my room. She thought they were there to help, not to violate our rights. After the police screamed at me and threatened to deport my mother, I finally gave in and told them that the ibuprofen was in the cabinet. The police officers looked at each other and started laughing. They were not convinced, so they began searching the entire trailer. Even after they found nothing, they still kept asking where the drugs were. I was baffled; I had already told them where they were. They told me that I needed to be at the courthouse the first thing the next morning to figure everything out. I agreed to show up.

The next day, I went to the police department with my mother because they still had several questions to ask us. They insisted that if I didn't show up, they would find me and deport my mother. As soon as we arrived, they took us to a small room on the second floor and closed all the doors. There was a bright light above us as I sat at a small table with my mother and the police officers sat across from us. They asked me a series of questions and kept trying to make me agree that I had been selling drugs. In reality, earlier, I had given some ibuprofen pills to my friend. Even though I told them the truth, they kept threatening me. They would say that they had already tested the "drugs" for purity and identification to make me say what they were, but I knew that I hadn't given any drugs to anyone besides some over-the-counter pills. They were trying to charge me with distributing a controlled substance. After nine hours of interrogation without any legal representation, my mother and I were terrified and exhausted. She insisted that I needed to tell the truth, or else they would send her and our family back to Mexico and leave my little brother and me in America alone.

It turned out that the friend I had given the ibuprofen pills was selling them and passing them off as a controlled substance. He had been dealing with financial hardships at that time. Once he got in trouble, he blamed it on me and said that I had given him the controlled substance. Since I was considered the original distributor, I got in trouble with the school and the law. To clarify, I did not distribute the ibuprofen to anyone; the only person I gave them to was my friend while we were at home. Furthermore, I had no clue about his plan to sell the pills at school. Even though parts of the situation were a misunderstanding, I still paid the consequences for my mistakes. Once the police had enough evidence to prosecute me for an imitation of a controlled substance, they began to ask me more questions about different things that were going on around town.

They were prodding into things that were completely unrelated to our case. I believe a rumor around town that somebody was selling real drugs, and they had assumed that I knew who it was. The police kept telling me that if I told them who was selling the drugs, they would drop all my charges. I did not know who they were talking about, and after a whole day of being interrogated, it felt like they were putting words in my mouth. It turned out that no charges were dropped, and I was prosecuted with a deferred judgment with a fifth-degree felony for imitating a controlled substance.

So, where did that leave me? A sixteen-year-old felon who had no idea what was going on. I was a criminal in the eyes of several people in my community because of what happened. If you live in a small rural town, you know that information tends to spread quickly like a virus. It was all over the papers to the point that it even got to the school, and they immediately expelled me. I was kicked out of the national honor society, kicked out of sports, and kicked out of school activities. I was not allowed on school grounds by any means and was disconnected from any contact with other students. I was vilified and dehumanized for what had happened. By the time I had received this news, I had accepted my fate and understood that my life would be a lot different. I started to think about my family and everything they had done for me and how I had hurt them with my actions. I had made a mistake; I'm only human. But had I made a mistake that was designed to happen? Had I fallen into the criminal justice system trap and become labeled a felon for giving ibuprofen pills to my friend?

As a teenage boy, I was unaware of what type of trouble I was in. I was scared and willing to do anything to cooperate, but the more I did, the worse it got. As a result, I lost everything I had ever worked for in life and high school. My motivation to pursue higher education was gone after the school and state decided to press charges. Eventually, I was convinced to plead guilty to a deferred judgment because we did not have the financial means to take the case to trial with a lawyer. The most important aspects of my life were put to the test. I showed everyone that I would not be a victim of a mistake and continued to pursue my educational and career goals.

Rather than succumbing to the status quo and becoming embittered and withdrawn, I overcame the odds with the help of my family and teachers. Even though I received two years of probation, I finished it within one year for unusual behavior. I was also very fortunate to have had white allies, such as my middle-school teacher, who still believed in me and knew that I was not a criminal but rather a product of a broken system. This incident is part of my story and is a lived experience that has positively impacted my life. I can proudly say that I have used this circumstance as a learning experience and a catalyst to further my educational goals. What I've described is a story that is not just unique to me, but to several First-Generation people across the world. What I have described is the American school-to-prison pipeline. Unfortunately, I know several other people or have heard of those who have fallen into similar circumstances. And for many, it ends right there—once they fall into the system and are labeled a felon and a criminal, there is no turning back. It's a cyclical mechanism put in place to keep people on the margins in America.

This story highlights how the system is indeed broken. However, it is not impossible to overcome. Overcoming the broken system should be the main priority for First-Generation students because we are often expected to fail. The standard is set up to get more Black and Brown bodies in prison rather than in school. At first, I was made to believe that I wasn't supposed to break out of that cycle; I was supposed to accept my new fate and believe that things were the way they were just because of my individual choices. However, we know this is false. We have learned that it's not only the combination of our personal decisions but also the context in which we make these choices that matter. While I do not discount autonomy in making decisions, I think it's crucial to understand the context in which people make decisions and the invisible forces that have pre-made those decisions. At the same time, I know that we can change the life course we are in, and the change begins within you. As New Americans, we must be visionary and never forget the struggles we have had to go through to become who we are today.

In this chapter, I will illustrate my story and personal experience of the school-to-prison pipeline. In the second part of this chapter, I will explain how several of the theories we have learned are evident in these injustices when it comes down to policy. I will be primarily using a health equity approach and examining criminal justice issues using epidemiology, social determinants of health, health policy, environmental health, and global health. At the end of this chapter, it will be apparent that there is a need to develop more comprehensive substance use policies, starting with eliminating the school-to-prison Pipeline.

The War on Drugs or the War on the Poor?

★ ★ ★

One of the best ways to highlight the close intersections between health and policy is by illustrating how the War on Drugs has had an enormous impact on generations of people and set up the infamous school-to-prison pipeline that is often talked about in America. In my opinion, the way the criminality of drugs has been weaponized to discriminate against racial minorities is one of the most significant health disparities we can see. Not only has it had damaging consequences in the lives of people directly impacted by the criminal justice system, but it has also decimated any opportunity, they have to build wealth for several generations.

The War on Drugs was a political tool given name and used by President Nixon in the seventies to unjustly and disproportionately incarcerate people of color.[1] Simultaneously, the US set up the school-to-prison pipeline, investing in prisons instead of in education.[2] Why? Because the people who were abusing their power valued profit over lives. Incriminating substance use was an easy way for them to institute more structural racist policies that targeted people who were non-white-appearing. Although we do know that white people represent the largest proportion of substance users in the population, they are not targeted in the same way by mainstream media and racist policies as Black and Brown people are.

There is a lot of controversy around substance use, especially when it comes down to the use, distribution, and manufacturing of these products. The political rhetoric around substance use took a dramatic change in the 1900s when the negative impacts of substance use were amplified and misrepresented in mainstream American culture.[2] The concept behind addiction and criminalization was presented as synonymous. As a result, this led to the mass incarceration of substance users, a disproportionate number of whom were Black and Brown people, as we have demonstrated. These incarcerated people were also human beings with families at home, yet tearing apart families and incriminating a whole generation of people was never considered by the powers that be.

During the conversations about and executions of policies that criminalized substance use, decision-makers neglected to consider the children and families who used substances. When the US's leadership decided to wage war against drugs in the 1970s, they mostly started a war against their people. Additionally, America was able to take advantage of another structural abuse of power to marginalize and decimate communities of color. All of this ultimately impacted the younger generations, such as the children of substance users. However, the criminal justice system did not stop there. They took a step further and even set up the expectation that these children were also criminals.

Substance Use as a "Crime"

★ ★ ★

Some people may wonder how and why the system is set up the way it is. This is a complicated answer, and it involves a long history of different policies that have been enacted in our country. In this section, I will highlight the specific policies and actions created to put us in the current situation we are in. I will highlight several of the ideas we have learned before and use public health sciences to demonstrate the different perspectives that are needed to understand how we are where we are today, what kind of things we can do, and how to use this information when advocating for the elimination of the school-to-prison-pipeline system in America.

As we have previously learned, one of the most critical social determinants of health is socioeconomic status (SES) because it can influence one's access to resources. The complexity behind low SES makes it challenging to pinpoint areas where specific interventions can prevent adverse health effects from substance use. However, we know that criminalizing substance use is just another way to systemically target communities of color, so one way is to start there. There is extensive research studying the negative health consequences of low SES among substance users and how it can lead to homelessness and eventually to incarceration. There are direct and indirect consequences of homelessness, incarceration, and low SES that lead to disease. Low SES can lead to imprisonment, and vice versa—it is a bidirectional pathway that creates a cycle. As this cycle progresses through life, it will also lead to limited access to resources, poor quality of care, and other social factors that influence disease. All these social factors are interconnected in a way that undoubtedly turns into a marginalization of groups of people that predispose them to disease. This indicates that public health interventions need to recognize at which stage it is most useful to intervene to disrupt the pathways between incarceration and adverse health outcomes.

Understanding these pathways is essential because it can result from the criminalization of substance use and its false association with people of color. The mainstream media makes it appear as if the largest substance users are people of color, and the police also overwhelmingly target these communities. Little do many of us know that the government has it wrong, and they have always had it wrong with substance use and incarceration. Several experts agree that the criminalization of substance use—the War on Drugs—was just a political tool to marginalize certain communities further.[3] Additionally, minority groups are being brought into contact with the school-to-prison system from an early age, resulting in inter-generational cycles of marginalization. In other words, it is all just a modern form of systemic racism. There is substantial literature suggesting that minorities report similar and sometimes even lower substance use than their white counterparts. Yet, they still have a lower SES status and experience much more significant adverse effects from the War on Drugs.

It is often presented that drug use will lead to incarceration, homelessness, and low SES. However, contrarily if we put all of these determinants of health in place, it is suggestive that these factors will lead to substance use instead. SES can predict high-risk behaviors that can lead to incarceration and poor health due to behavioral norms and a lack of access to resources. To adequately address the determinants of disease among substance users, it is essential to approach the issue through a top-down and bottom-up policy lens. Individual behavioral interventions will not sufficiently address drug health disparities; instead, we need to adopt a more holistic approach, such as improving economic stability, decreasing rates of recidivism and pursuing drug policy reform.

Substance use has not always been classified as a crime. If we look throughout humans' history, substances have played an es-sential role in many populations throughout various geographical locations. For example, indigenous groups, such as the Aztecs, used mushrooms and other hallucinogens for ritual and spiritual healing ceremonies, a critical part of their culture.[4] Even until this day, there are Native American groups who use substances like tobacco for spiritual traditions. Substance use has a long-embedded history in our world.

Furthermore, substances have a legitimate purpose in medi-cal use to sustain the population's health and welfare. One recent example is increasing evidence regarding the medical and health benefits of the use of marijuana in end-stage diseases.[5] There are also trials in a better understanding of how doctors can use ketamine therapeutically for psychiatric conditions.[6]

Today it remains that there are apparent issues when it comes to substance use and addiction. Many people argue that addiction is a choice, but there is broad scientific consensus that it is pathologic in nature. The main problem is that drug addiction is perceived in America as a crime and stigmatized, instead of seen as an issue of health. In other words, a person who is a substance abuser (or someone who suffers from addiction) is treated like a criminal by America, rather than as a patient who needs health care. Instead, in our country, people usually end up back in prison, where they continue to face a vicious marginalization cycle rather than getting the treatment they need.

Today, the War on Drugs and the school-to-prison pipeline involve substance criminalization as a method to incarcerate non-white people. All this to say that the regulation of substance use with policy and law is complicated. Our current approach is not cutting it, and it will not unless we institute major reform of the criminal justice system.

A History of Substance Regulation in the US

★ ★ ★

It is imperative to look at drug use in the Americas and its impact on health through a sociohistorical perspective to get an idea of the specific trends and patterns of drug policy. Substance regulation in the US has dramatically shifted from none to little control to the heavily criminalized penalties today for use and distribution. As I described above, some of the country's earliest drug use accounts were during precolonial times among Indigenous tribes. In the 1600s, colonial law required farmers to grow hemp for industrial purposes, which continued until the 1900s. During this time, there were instances where the prohibition of alcohol was not successfully implemented. After the American Medical Association was founded, the patent medicine industry started to rise; opiates, cocaine, and other drugs were sold over the counter but were primarily used for medicinal purposes. It wasn't until the 1900s where drug policies started to change dramatically in the US.

Drug criminalization and heavy regulation began with the Pure Food and Drug Act in 1906, where pharmaceutical products were required to be labeled with ingredients. Since drugs were widely accessible to the public, dependence on opium and cocaine became a severe social problem, so the government responded with the Harrison Act of 1914, which mandated that some drugs be prescribed only by a physician. The criminal justice system and the government began an anti-drug momentum, which eventually led to the Nixon administration declaring the War on Drugs in 1971, which composed of a wide variety of drug control policies such as the Rockefeller Drug Laws and the Controlled Substance Act. These new laws and policies led to a substantial increase in the prison population nationwide. Overall, these regulations' impacts are said to produce adverse health outcomes in specific communities, especially children. These laws significantly impact children due to high incarceration rates, where there is an increase in separation from their parental guardians, which can have detrimental effects. Between 1991 and 2007, the number of parents in prison increased by 70 percent, resulting in many mechanics in children who experienced the exposure of having a parent in jail, which increased the risk of adverse outcomes.[7] Some studies suggest that drug policy impacts children's overall wellness, which leads to the health disparities we see today. These adverse outcomes in our children demonstrate the need for policy reform to address the issues that face our country and the world.

Substance Use in the Context of Global Health Policy
★ ★ ★

When it comes to how substances are handled globally, there is a broad spectrum of substance use acceptance, prohibition, and regulation. On one end, there is the US, where all drugs are regulated or prohibited with the Controlled Substance Act. There are different methods through which the US controls drug use, and these are usually carried out through the criminal justice system, as we have learned previously. On the other hand, a country like the Netherlands has legalized and decriminalized many of the illegal drugs here in the US. The principal drug control regulations in the Netherlands are education and prevention while allowing their citizens to have the free will to choose to do drugs or not. Instead of punishing people using drugs, the country gives them options for safe substance use.

Since the US is a global superpower and has a lot of influence on other countries, more countries criminalize substance users rather than decriminalize them. Global drug prohibition is a worldwide system structured by a series of international treaties supervised by the United Nations. Many of these treaties allow each country to prohibit controlled substances by police and military forces, giving countries the potential to use policy to increase criminalization and incarceration. When it comes to substance regulation and prohibition around the world, there should also be careful considerations of the cultural contexts that they are being implemented in. Global drug prohibition is an invisible system that has been kept undercover, taking over drug policies in multiple countries. However, since each country regulates its drug policies, drug prohibition becomes a spectrum where substance use regulation is relative to the nation that governs it. Global drug prohibition has minimal benefits and can result in multiple crises that have to be examined.

With the US playing a significant role in influencing other countries to follow similar drug control tactics through criminalization, governments will receive additional power to use the police and military, not just to prohibit drugs but also in unjustifiable seizures in non-drug operations. In contrast, these countries' people will be unjustly impacted by these laws and policies that further widen health disparity gaps for racial minority groups. The militarization of the War on Drugs is a perfect recipe for violence due to the lack of supervision and accountability. One brief modern-day example of this war's failure is Philippine President Rodrigo Duerte, who was elected in 2016. As a result of his "war on drugs" there have been at least 12,000 Filipinos dead through the use of extrajudicial violence against the poor.[8]

The uses of anti-drug messages and drug demonization are beneficial to the school-to-prison pipeline because it gives governments the power to socialize their citizens to develop individual drug attitudes that can result in stigma, instilling fear in people. When governments use mainstream media to correlate drugs with other types of crimes, it generates a social problem that worsens drug issues. Drug prohibition also can unite political opponents for the "common good," where political leaders share similar values like governmental police powers to enact harsh political movements. Some of these movements have resulted in punitive anti-drug laws, expanding prisons, and transforming a war on drugs into a business.

While the drug control business benefits governmental powers, it has impacted multiple communities in negative ways. Some major turning points that drug prohibition continues to struggle with are harm reduction movements and anti-punitive drug policy reforms. Harm reduction methods in public health initiatives are useful because they reduce the negative impacts of drugs in general. One of the most recent harm reduction approaches has been reducing the spread of HIV/ AIDS and hepatitis through syringe exchange programs.[9] These types of programs fit into the harm reduction model because it acknowledges that no matter what we do, people are still going to use substances. Therefore, it is our responsibility as a society to ensure that we provide all the information available for people to make important decisions about their health and provide them with the tools to keep them safe.

The harm reduction model essentially posits that people will engage in these behaviors anyway, so it makes sense that we try to make the situation safe and use it as the perfect opportunity to provide other services to help them with substance use and addiction. Harm reduction encourages policymakers to shift drug policies away from punishment, coercion, and repression toward tolerance, regulation, and public health. Drug prohibition will not be able to compete with these new drug movements. Also, it is not a matter of whether drug prohibition will end; it is a matter of when it will end.

In this chapter, I have illustrated my encounter with the school-to-prison pipeline in the US as a first generation student of color. We have translated theory into practice by examining different examples of how the issue can be understood using a public health equity lens. We have also highlighted the various policies that have resulted in the situation we find ourselves in today. We have learned that if sound anti-punitive drug reforms do not replace current drug policies, children and young people will continue to be harmed. As incoming doctors and professionals, we must be aware of the school-to-prison pipeline to identify better ways to oppose it and instead strengthen other beneficial pipeline programs that puts on a path to health equity.

If you or know someone who has been caught up in the criminal justice system just as I was, just remember don't let it define you. And for those who have not been caught up, don't fall for the trap. The system was designed like this for a purpose, and now it's time we dismantle it and redesign it by learning the history of how it came to be, what kind of new identity we must uphold and value as a country, and how we can cross-collaborate across the different systems to eliminate the school-to-prison pipeline.

REFERENCES

1. Alexander M. The New Jim Crow: Mass Incarceration in the Age of Colorblindness. *New Press.* 2010.
2. Mallett CA. The School-to-Prison Pipeline: A Critical Review of the Punitive Paradigm Shift. *Child & adolescent social work journal.* 2015;33(1):15-24.
3. Eldredge DC. *Ending the war on drugs: a solution for America.* 1st North American ed. Lanham, Md;Bridgehampton, N.Y;: Bridge Works Pub. Co; 1998.
4. Schultes RE. Teonanacatl: The Narcotic Mushroom of the Aztecs. *American anthropologist.* 1940;42(3):429-443.
5. Birdsall SM, Birdsall TC, Tims LA. The Use of Medical Marijuana in Cancer. *Curr Oncol Rep.* 2016;18(7):40.
6. Zanos P, Moaddel R, Morris PJ, et al. Ketamine and Ketamine Metabolite Pharmacology: Insights into Therapeutic Mechanisms. *Pharmacol Rev.* 2018;70(3):621-660.
7. Turney K. The Unequal Consequences of Mass Incarceration for Children. *Demography.* 2017;54(1):361-389.
8. Philipines' Duterte apologizes to Obama, makes friends with Netanyahu. *UPI Top World News - Newspaper Article* 2018.
9. Vearrier L. The value of harm reduction for injection drug use: A clinical and public health ethics analysis. *Dis Mon.* 2019;65(5):119-141.

10

The New American Dream

A call to action for the New American to respond to and transform the structural inequities in our country

"If not us, who? If not now, when?"

—*John F. Kennedy, Thirty-Fifth President of the United States*

At the end of my first week in college, I felt empowered and invincible. I was also a bit nervous, as this was my first time away from home. Suddenly, my cell phone rang, and an unknown number from Mexico appeared on the screen. Who could it be? When I answered my phone, I heard a boy screaming, "WILLIAM, PLEASE WAIT!" He was breathing heavily and quickly, perhaps running for his life. I kept asking for his name. Who was he? How did he know me? It was as if I could hear the pounding of his heart. I hung up. Right away, the same number called my phone. I knew the situation was severe. After I answered, my paternal grandmother informed me that the boy who was calling was my cousin. There had been a family emergency, and without giving me any reasons why, she urged me to ask my mother to immediately call her back. Two hours later, my mom knocked on my dorm room door. Devastated, she told me, "Your father is dead. He was murdered in Mexico." After my father's passing, I began to view myself as a New American in pursuit of the New American Dream.

After the passing of my father, there were many challenges I faced including suicidal thoughts with severe depression, nonetheless I was able to persist through this challenge. Having to deal with so much grief, I felt the need to drop out of college, but instead, I decided to stay in school and keep pursuing my education. I remained in school because as I learned more about what it meant to be a New American, I also learned that my journey was in honor of my father and family who have sacrificed so much for me to allow me to get to where I was. It would be a shame if I just dropped it all and stopped pursuing my dreams. Instead, I used this negative situation to fuel a positive mindset and a new identity to pursue opportunities that would teach me how I can begin to address the issues that led to the death of my father.

How I Became a New American and My Pursuit of the New American Dream
★ ★ ★

The New American is a unique perspective in which people from American society have the opportunity to recognize where we have come from, appreciate our diversity, and strive for more equity. The New American is an identity you can adopt and call yourself if you believe in what I will share in this chapter. The goal of this chapter is to provide insight and guidance into the new values and aspirations that America should uphold while reconciling and recognizing the historical traumas we have endured. As I proclaim this new identity, I am acknowledging those who have walked before me and my ancestors, who lived in peace with our earth. Being a New American means holding on to your roots of where you came from, holding on to what you stand for, holding on to what you believe in, and holding on to all the experiences that you've had in your life that have allowed you to become the persons you are today.

A New American is like a tree, having strong roots to your background, while sprouting in different directions as you discover your passion and your purpose of how you can serve human society toward never-ending horizons. Being a New American means you have a new framework and a new mentality built on the foundations of we have learned in the first part of the book, which has allowed us to describe the mechanisms of inequities in our society. Considering the amount of harm that has been done on the American continent, as well as the rest of the world, it is our moral duty and obligation as New Americans to uphold our principles to the greatest extent. Only then will we be able to see true peace, prosperity, and healing for all humans, creating the New American Dream.

Everyone can be a new American. You just have to think like a New American. One simple way to achieve this is by acknowledging that the founders of our nation are not the people we've been taught to think they are, while at the same time understanding the legacies that they have left behind. While acknowledging of the great they have done, it is also just as important to acknowledge the things they did wrong so we can begin to repair this imperfect system they left behind for us. It is essential to know this in order to work together and begin dismantling systemic forms of oppression. As New Americans, we can create healing spaces and change institutionalized policies that impact our daily lives. The pursuit of health equity is the New American's goal, and only until we are able to achieve health equity will we also be able to embark on an equitable pursuit of life, liberty, and justice.

My New America Dream began when I was born in East Los Angeles, California. After living there for a couple of years, we moved to the mountains in Colorado. As a young boy, I always believed in the American Dream. However, after I went through several of the hardships I have detailed in previous chapters, the American Dream for me was more like an American nightmare. The reality of being the son of two Mexican immigrants and experiencing the obstacles that many First-Generation Americans undergo was a common theme throughout my life. I felt like I was never able to fit in anywhere I went. To my American friends, I wasn't American enough, and to my Mexican friends, I wasn't Mexican enough. On top of identity confusion, I also faced financial and personal difficulties. But instead of succumbing to those challenges, I began to draw from them and decided to overcome my adversity with the diverse skills I could bring to the table.

As I grew older in rural Colorado, I could not help but notice the structural inequities that persisted in my community. I just didn't know what to call them until I was able to learn from them. I just thought they happened because that was how things were supposed to be. I never questioned why all of the Mexicans lived in a trailer park in the outskirts of the town with no access to transportation or even sidewalks. Injustices permeated all aspects of my life, evidenced by the fact that my neighborhood had one of the highest high-school dropout, teen pregnancy, and poverty rates in state. My family was denied essential life resources, such as health care and job security. It was challenging to maintain a healthy life and remain positive in the face of many of the adversities we faced.

My family always valued education, but as undocumented immigrants, many educational opportunities were not available to them. Not only were there significant financial barriers but there were also language and cultural barriers. Since my family could not make enough money to live in town, I grew up in a trailer park community about three miles from the actual town, making it challenging for us to access resources. Since we were a lower-income family, I worked throughout high school to help my mother in the hotel housekeeping industry, as well as a lifeguard to help make ends meet. When my encounter with the police occurred in high school, instead of becoming bitter, with the help of a mentor, I learned I was able to turn things into a learning experience. I assumed complete responsibility for my actions and performed more than 150 hours of community service with a wide variety of organizations. During high school, I enrolled in Colorado's dual enrollment program, which allowed students to earn college credits while completing high-school requirements, which I did at the same time as my lifeguard job at the local community pool. By the time I graduated high school, I was able to start with thirty-three college credits. Little did I know that I was just starting to begin my journey as a New American.

It was a miracle that I graduated from high school, especially since many doubted me after I got in trouble with the criminal justice system at the age of sixteen. No one expected me to make such a drastic turnaround. Not even my father, who had to return to Mexico to take care of his family. He was surprised that I graduated but was overall proud and excited to get back to the US so he can support me on the journey that no one else in my family had ever pursued. Although we had more economic opportunities in the US than in Mexico, we still lived in poverty as undocumented citizens. It was difficult for my family to stay together, given that my father was always gone and working three jobs, while my mother was working a full-time job.

Since my dad was hardly around, my mom could not take the pain of not seeing him and resulted in a divorce. After this, I split my time between both parents, living with my mother during the weekdays and with my father during the weekends. All of these combined experiences had a profound impact on my upbringing because I believed I would not get a chance to pursue the American Dream because I thought my family was broken. Since I was growing up in a family that had had to work most of their lives instead of study for survival, graduating from high school or going to college was never something I imagined myself doing. Thankfully, with the support of some high school counselors, teachers and mentors I was not only able to persist but do well academically in high school, enough to graduate with scholarship support to pursue college.

At the very beginning of my college experience, I was curious about being involved in community- and campus-wide health equity activities and programs. Through my involvement in these different programs, I evolved into the person I am today. All this to say that by being involved in the things you care about and are interested in, you will be able to find your passion. Although making a living is important, being happy with what you do is arguably more important. Through my experiences, I became certain that I would be able to approach all challenges in life with my thirst for knowledge and self-improvement, with utmost vigilance, and a strong desire to make the world a better place through service to the community.

The Vision of the New American Generation

★ ★ ★

Being a New American is a direct result of the intersectional identities and experiences we have in our lives. To be a New American means consistently rediscovering your roots and placing the highest value on your heritage, even when society tells you otherwise. For example, at first, I found being labeled an immigrant hurtful; after all, I was living on my ancestors' land. While I grew up identifying as an American, I was also labeled Latino, which confused my identity. Eventually, I settled for half-American and half-Latino. I resisted, and since then, I have discovered that as a New American, I am not two halves of a person, but rather a full human being, wholly and profoundly worthy of acceptance and respect.

Being the first in your family to pursue a profession and educational achievements will make your family and community proud.

More important, pursuing an education as a New American gives you the knowledge and know-how to impact people's lives in meaningful ways. New Americans are the new role models that I hope we can create for our future generations, as well as even in our current one. For example, I want to be a role model for my younger brother, peers, and other immigrants who aspire to become health professionals themselves. My father taught me that although changing the world is difficult, the transformation starts with one who is willing to be the fundamental change.

A New American is not only an identity one can have but also a mentality that does not succumb to the status quo. This means that just because things are the way they are does not mean that this is how they are supposed to be or that we should just accept things the way they are. I am not saying that we should not celebrate our successes; rather, I'm saying that we still have a lot of work to do, as evidence by this book. A New American will use personal, educational, and professional experiences to shape the way they perceive the world and ultimately motivate others during times of uncertainty by giving support and compassion.

Intersectoral and Multidisciplinary Community Collaboration for Health Equity

A guide to the interdisciplinary approach to public health equity

"I'm for truth, no matter who tells it. I'm for justice, no matter who it is for or against. I'm a human being first and foremost, and as such I'm for whoever and whatever benefits humanity as a whole."

—Malcolm X

The Use of Intersectoral Collaboration with Health in All Policies

★ ★ ★

The first time I was able to witness intersectoral and multidisciplinary collaboration was when I was a student participating in the Rocky Mountain Public Health Case Competition. This case competition is hosted by the Colorado School of Public Health and helps students across varied disciplines work together in teams to design and create innovative solutions to real-world health problems. My team consisted of a two public health students, a nursing student, a pharmacy student, and myself representing as a medical student. Our topic was addressing health issues that arise in a community following a natural disaster, specifically a flood. When designing our solution, we realized how our multi-disciplinary approach was helpful because we all had unique perspectives and experiences to bring.

As we continued to design our solution we discovered that collaborating with other sectors such as the educational, housing, food, and public safety sectors were all an integral part of our solution. The educational sector was important to engage because they were the primary people who was responsible for design information materials about the spread of disease in water during floods. The housing sector became important because they were responsible for creating temporary housing for those who had their homes destroyed by the flood. The food sector was important because many of the crops were destroyed by the floods and they needed to ensure they can still supply enough food for people who lost everything. The public safety sector was important because they had to ensure displaced people were safe and that violence in times of stress were minimized. All of these sectors were just as important as the health care system in ensuring all the citizenry remained healthy during a flooding disaster.

One of the most obvious ways in which the US can begin to examine health disparities and set systems in place to address them is through intersectoral collaboration. Intersectoral collaboration is the collective actions involving more than one specialized agency or group, performing different roles for a common purpose and goal.[1] Using an interdisciplinary approach to implement public health equity interventions is one of the main ways to address the health inequities we have discussed in this book. Public health equity is the concept of critically analyzing the health disparities we see rooted in historical abuses of power that cause adverse health outcomes at the population level.[2] And by an interdisciplinary approach, I mean using a system that combines different professions (i.e., doctors, lawyers, teachers) and sectors (private, public, nonprofit) to work together to come up with solutions instead of working in silos. Intersectoral collaboration is a promising approach that we can utilize to move American society to a place where we can all be proud, happy, and safe to call home. By understanding how different systems and disciplines work together, we can leverage each other's strengths to improve humanity.

Adopting an intersectoral approach enables people to collaborate and share ideas while not replicating efforts or to reinvent the wheel if you will. The utility behind working in a concerted and organized effort allows for the greatest systemic and sustainable solutions. Intersectoral collaboration all starts with sharing perspectives and practices and acknowledging that everything is interconnected and that everything impacts others directly and indirectly. Through an intersectoral approach, we can address systemic issues that contribute to adverse health outcomes. For example, a critical health systemic issue is affordable access to health care itself, which creates even larger health issues since it further perpetuates inequality.

In the US, health is not valued as a human right. Instead, the healthcare system operates more like a business than an institution of health given the limited resources that exists. In most cases, if you do not have any money or insurance, you are unable to get care without getting into crippling medical debt. While it is essential to have a cost-effective health system, human lives are far more important than profit margins. Through empowering other sectors to consider health, it will increase the ability for us to respond and address health concerns. Rather than relying on politicians' opinions on the best way to provide care, rigorous research rooted in strong scientific methods should be used to discern theories and practices that can inform health laws. Only then can we leverage evidence-based practices to fully optimize the limited resources we have on this earth.

With intersectoral collaboration, we can evaluate and assess the impact of certain policies and systems on the human condition. It is essential to determine which systems are not working and fulfilling the purpose they are supposed to or originally intended to do. With an intersectoral approach, it will be much more efficient to ensure equitable healthcare approaches. Instead of the traditional way, we leave the healthcare system to only deal with health issues. In the next part of this chapter, I will highlight some specific examples of certain ways society can use intersectoral collaboration to achieve health equity.

One of the primary ways intersectoral collaboration has come to be practiced is through the use of Health in All Policies (HiAP), a framework that examines the health impacts of policies and practices outside of the traditional healthcare system.[1] Whether we are talking about the criminal justice system, the educational system, or the transportation system, they all have the potential to cause health effects on people through the social determinants of health (SDOH) as previously discussed. An HiAP approach is an emerging practice that health agencies are starting to use to collaborate with others outside of health and include them in the decision-making process to ensure that health is always a consideration when trying to implement new policies or programs. Not only is HiAP the cornerstone of the success of intersectoral collaboration, but it is also a tool that we can use in life to examine and implement systemic change while elevating voices that are not traditionally heard. The emerging practice of HiAP has been successfully implemented at several government levels, including the local, state, and federal settings. I have personally been able to study local and state examples of HiAP, mostly being involved in local settings.[3] In my experience, I have learned that working with the local community can be an effective way to enact changes with visible impacts using an HiAP approach.

Practical Steps and Recommendations to Achieve Intersectoral Collaboration

★ ★ ★

The most practical way to start addressing outcomes is by identifying what these health outcomes are and learning more about what can cause or contribute to the same outcomes. One example of a health outcome is increased morbidity rates of metabolic syndrome, including several different health conditions, such as coronary artery disease, diabetes, high blood pressure, and waist circumference. One of the main issues among populations with increased metabolic syndrome rates is increased exposure and consumption to increased unhealthy food.[4] However, what we don't look at is the availability of food quality and access. The places where we see the highest metabolic syndrome rates are also in the same neighborhoods that lack healthy food access in their community. Food access, racial composition, and income can be directly linked to health outcomes based on neighborhood environment demographics and as such should be the primary factors that must be addressed. Neighborhoods where food is not equitably accessible and available, have continued to be relatively segregated according to household income level and race and therefore increasing racial and socioeconomic diversity should be a priority. There are social structures in place that have led to race and class segregation by neighborhood that must be acknowledged and reversed. Investing in communities that have been previously redlined must be prioritized. By emphasizing these different initiatives, we can begin to address the issues that contributes significantly to poor food environments within marginalized communities.

Because food access is linked to poverty, to improve health outcomes in these neighborhoods and eliminate health inequities, there has to be a focus on the political and social processes that lead to the deconcentrating of poverty within these neighborhoods and in the broader economy. One way to change and improve the health of these lower-income neighborhoods with a high concentration of poverty is to incorporate accessible, affordable public transportation as well as diverse land use that provides employment opportunities. We can do this by providing access to public transportation funds from taxes and providing community members access to employment and education opportunities that can help break them out of the poverty cycle. Suppose we invest less in expanding the highways for private vehicles and put more money into building robust public transportation that allows people to easily get to their jobs. In this case, there will also be fewer car accidents and less pollution, which will continue to contribute to healthier communities. We will help eliminate health inequalities for people of color in lower socioeconomic neighborhoods by doing these different types of initiatives.

Health and allied social welfare professionals must use their expertise to guide government officials in decision-making by providing rigorous evidence-based interventions when available and also highlighting opportunities to collaborate with other sectors of society. Whenever there is a lack of evidence, it does not mean that the action should not be considered. In other words, lack of evidence should not be the end all for not enacting any changes and keeping things in the status quo. Nonetheless, all intersectoral designs must be sustainable and systematically safeguard the health of the community in the interest of the greater good. By focusing on the external environment beyond the individual, we can supplement individual behavioral interventions and recognize the limitations that each system has that prevents them from addressing the issues at hand individually.

Combatting Racism with an Intersectoral Approach

★ ★ ★

There is a need for more robust intersectoral and multidisciplinary initiatives to address racism in our society since it has been built into every aspect of America's fabric. Racism has been impacting our nation's health for hundreds of years, and we have yet to explore solutions and how we can begin to let our country heal from our traumatic past. Instead of solely relying on individual interventions to address America's historical racism, we should focus our efforts on using medical-legal responses that target structural pathologies that cause inequities in health. One example of this response is through the use of Medical-Legal Partnerships (MLPs) which is a multidisciplinary team that works together to address medical and social/legal problems that have an impact on overall health.

Racism is a public health problem and a private concern because it creates social and environmental determinants that are harmful to our health or the health of people we know. Much of the dispute around addressing racism using the model of public health and medicine is understanding what mechanisms drive the damages and how it causes these physical, social, and economic outcomes. Most claims to counter the impact of racism in health are grounded in denial or lack of personal responsibility and compassion for the human condition. Although there has been an increased awareness of the problem, and it does not seem like it will be going away anytime soon. Integrating anti-racists practices in all sectors of society is crucial to setting us up to change behaviors on a mass scale.

Several of the interventions proposed in this chapter can be implemented at many levels including the individual, community and societal level. Intersectoral collaboration will make it easier to implement restorative justice policies and increase awareness of the issues that impact minority populations. Ultimately, there has to be a more significant and larger share of responsibility from the population and a greater focus on addressing factors beyond the sphere of personal responsibility, as we have done for several years. In all cases, achieving these objectives bring are legitimate government concerns which include national security, mitigating the economic impacts of racism, and ultimately protecting citizens' health, particularly our future generations.

I propose that we become a society that is not colorblind, where we acknowledge our color and the real power associated with it. Simultaneously, we must concede that we are all one human race according to scientific, legal, ethical, and moral constructs. As humans, we must unite in these utmost essential aspects of our lives and recognize the need for better, robust intersectoral collaboration that protects populations who have been historically marginalized by the abuse of power. Some believe that they are not responsible for their ancestors' actions, and their logic might be understandable. But the fact is that some people are still reaping the benefits from a system that continues to oppress others, and our call as fellow humans is for recognition, a sense of responsibility, allyship, and collaboration. Most of us, including me, believe we are responsible for the next generation of humans who will walk our earth. If that's the case, we all need to do our part to acknowledge that our current situation is not something we can continue to stand for.

Using intersectoral collaboration allows us to unite from all different areas of society. The next generation must use the tools set by this chapter to identify the root causes of the issues we face in our communities. From margin to margin, we must come together and understand that we are all connected stakeholders in our common human experience. Racism is like a disease in the human condition that has impacts in the collective health of generations. We must rise and defend ourselves from threats to human health and prosperity and challenge these ideas. Addressing conditions such as racism using public health equity and medicine models is imperative for our continued existence and success. One way we can all unite besides practicing intersectoral collaboration is if we can identify as one—there is power in the numbers. To many people, the American

Dream is dead, and to several others, it was just an illusion and that is why we must re-create a new vision and dream. After reading this book, I hope that the next incoming First-Generation students and professionals across the spectrum can take on a renewed American identity in the pursuit of the New American Dream.

REFERENCES

1. Shankardass K, Muntaner C, Kokkinen L, et al. The implementation of Health in All Policies initiatives: a systems framework for government action. Health Res Policy Syst. 2018;16(1):26.
2. Hearld LR, Alexander JA, Wolf LJ, Shi Y. The perceived importance of intersectoral collaboration by health care alliances. J Community Psychol. 2019;47(4):856-868.
3. Mundo W, Manetta P, Fort MP, Sauaia A. A Qualitative Study of Health in All Policies at the Local Level. INQUIRY: The Journal of Health Care Organization, Provision, and Financing. 2019;56:004695801987415.
4. Jahangiry L, Farhangi MA, Rezaei F. Framingham risk score for estimation of 10-years of cardiovascular diseases risk in patients with metabolic syndrome. J Health Popul Nutr. 2017;36(1):36.

Combating the Structural Pathologies of Health

The role of New Americans in eliminating systemic inequities

> *"Laws are not science; they are normative ideology and are thus tightly tied to power. Biomedicine and public health, though also vulnerable to being deformed by ideology, serve different imperatives, and ask different questions. They do not ask whether an event or a process violates an existing rule; they ask whether that event or process has ill effects on a patient or a population."*
>
> —*Paul Farmer,* Pathologies of Power: Health, Human Rights, and the New War on the Poor

The New American Way of Addressing Systemic Inequities in the US

★ ★ ★

During the Winter of 2014, I had the opportunity to engage in service learning focused on addressing homelessness, injection drug use, and spread of blood borne infections in San Francisco, California. This experience was the first time I was able to begin to combat structural pathologies of health. To get a better understanding of how I was addressing these systemic inequities, I first want to explain what the structural pathology is I was trying to address. The structural pathology in this case is how current policies and laws lead to criminalization of people who use illicit drugs barriers for employment which then leads to disproportionate rate of unstable housing conditions and increased risk of infections.

Given the complex nature of this structural pathology, there are many different levels in which I was able to intervene. The most memorable encounter I had was when I was working with the San Francisco Aids Foundation where I helped assemble sanitary injection kits to reduce the re-use of needles which could be potentially contaminated. When we handed these kits out, many people were perplexed and thought that we were enabling people to use drugs. However, in reality this is far from the truth and instead just creating conditions that are safe. If people are going to be using injection drugs, our goal should be to make sure it is as safe as possible while at the same time offering ancillary services such as substance use counseling, treatment, and social support services. Throwing people in prison will not solve this issue. We took things a step further and worked with organizations that worked with communities experiencing homelessness and offered support by helping provide a meal and most importantly offering time and compassion through hearing their stories and acknowledging their struggles. These encounters are all different ways in which it is possible to address some of the structural pathologies of health.

Throughout the book, I have highlighted how structural pathologies can cause adverse health outcomes at the population and individual level if not addressed with a pragmatic approach. Structural pathologies refer to the adverse social conditions embedded into systems that lead to inequities in health. Another example of a structural health inequity is how the practice of discriminatory hiring practices can lead to poor health outcomes. For example, discriminatory hiring practices puts people with non-American sounding names at a greater disadvantage at getting a job and as a result will lead to a hard time finding employment. Without employment, there is no money to buy healthy food. Without healthy food, you are more likely to be at higher risk of chronic medical diseases. However, without any employment, chances are that you also don't have insurance and as a result are unable to get care or buy medications for your health. All resulting in adverse health outcomes compared to those who benefit from discriminatory hiring practices.

Whether we are talking about differences in primary outcomes such as mortality or rates of disease, or secondary effects such as socioeconomic status, these outcomes are connected by complex social mechanisms that influence each other in a bidirectional manner. Meaning, we can't study these different topics in isolation and require an intersectoral approach. This chapter will discuss some final thoughts regarding the role of New Americans and how we can address health inequities and present a quick summary of the different topics covered throughout the book.

New Americans have unique perspectives and ideas to challenge the structural pathologies in our country. Given the diverse background of New Americans, we all have different views and best practices to provide solutions for intractable health problems that one group would not be able to do on their own. A diverse community requires diverse solutions, and as New Americans armed with the knowledge in this book, we are now prepared to advocate for these solutions that will create positive change. The American workforce is continuing to age and will have to be replaced soon; therefore, we will need to replenish a new workforce with fresh ideas. Since New Americans come from all parts of the world, we can indeed be a country that is the most diverse place in the world and embraces these differences as strengths.

Addressing structural pathologies will always be a moving target that evolves along with current society's norms, practices, and values. And as such, it is crucial to remain vigilant. Since structural pathologies are related to the social context, we must remain aware of new ways that they form or go unnoticed. This moving target will cause accompanying structural pathologies difficult to trace or pin point. For example, we went from colonization and slavery to Jim Crow laws and then the civil rights movement and on to mass incarceration, which is the industrial prison complex in place that targets ethnic and racial minorities. From the outside, these all seem like different ideas, but in reality it has been the same ideology just evolved to fit the current historical context. We are currently dealing with one of the most challenging years that will go down in history, and we need New Americans to step up to bring our country back. COVID-19 has struck and exacerbated our current social conditions, yet it also revealed many of them.

Using public health sciences and critical race and ethnic studies, we can understand diversity, inclusion, social change, and social responsibility in the context of a local, national, and global multicultural society. We must analyze health from a ten-thousand-foot view, rather than through individual, simplified narratives. With that in mind, as New Americans, we must plan to use holistic social and environmental justice approaches to serve diverse communities. Using environmental justice, we can move towards a society that provides fair treatment with respect to the development, implementation, and enforcement of environmental laws, regulations, and policies. This approach is integral to improve and maintain a clean and healthful environment, especially for those diverse populations who have traditionally been exposed to sources of pollution that are known to cause adverse health outcomes.

At the core of effectively serving a diverse population, it is essential to collaborate across our differences to meet the community's needs. We must focus on using our differences as strengths to develop interdependence where community members can become empowered leaders themselves to make changes or speak out when their demands are not being met. Building relationships in the community is imperative to establish an effective communication method. Culturally responsive engagement includes, but is not limited to, language access services and improving health literacy. Using collaborative and reciprocal relationships will allow us to have an asset-based community building rather than building through a community deficit model.

Addressing Systemic Inequities Inside the Healthcare Field and Out

★ ★ ★

When ensuring that all people have the highest quality of health care, it is crucial to understand that health does not just happen inside the clinic or the hospital. Growing up with Indigenous roots, I have put the highest value on my heritage that has taught me that health encompasses physical, emotional, mental, and spiritual well-being. On top of this, we need to raise awareness about a new approach at looking at health which includes biological, environmental, and complex social processes. Structural pathologies of health will take more than the health-care system and health professionals to address. For example, teachers can also begin to create significant change in their classrooms. Teachers are fundamental in children's upbringing. They spend a lot of time with children and can choose what they teach their students. Teachers become the backbone to society and have the capability of serving as role models, offer guidance, and dedication to further develop our society and economy.

One example of how a teacher can intervene at the level of education to help address these pathologies is by being aware of the school-to-prison pipeline and advocating for minority students to earn an education. I remember when I had some of my teachers support me when I was a teenager. My teachers and mentors helped me believe that I could still pursue education even though I had gotten pushed out of high school. One particular teacher, a member of the Upward Bound TRiO program, which encouraged me to take an exam that allowed me to earn college credit in high school. I also had an incredible mentor who took me on my first and only college visit to the University of Colorado Denver. Because of the support of teachers in my life, now I will one day be able to save lives as a physician and contribute to the advancement of all human beings. When we engage in other areas that also contribute to health, we can create a holistic approach to creating a healthier future for all. On top of engaging the educational sector, we also have to engage in other sectors with equal passion and dedication.

One particular sector that needs further work is the political sector. Policymakers, whether we are talking about lawyers or politicians, also need to step it up and fully engage with their systems. Lawmakers can define the realities in which we live and set many standards. We need increased representation in policymaking to enact truthful and meaningful change in diverse communities. Consequently, I recommend that policymakers improve social and living conditions for all Americans, particularly those who are low-income and have a racial/ethnic background. Policymakers should focus on narrowing the income gap by providing fair and equitable opportunities for upward generational mobility. Another recommendation would be focusing on improving medical care both in terms of access and quality. The Affordable Care Act was a step in the right direction; however, it still has several gaps and therefore a Medicare for all program would be superior. There is a demand for training culturally responsive health professionals. Policymakers need to address the current lack of incentives for providers to practice in underserved areas with the greatest need. This is not to say that those in most of the population do not experience things like poverty, lack of care, discrimination, or police brutality, but they don't experience to the same extent and scale that minorities do.

Politically, this has been hard to achieve because of party loyalist. As a result, politics in the US has become polarized and split into two groups, the Democrats and the Republicans. Both parties have failed several generations. We pride ourselves so much on "freedom" and cast shame on countries that are one-party states. But how much more free are we really by having two options instead of one, where both options still don't provide what we truly need? This country is so two-party obsessed that reforming our democratic participation is the best thing to do. New Americans who want progressive change will have to do away from old politics. We must detach ourselves from corrupt political parties who prefer profits over lives. We must not be party loyalists. Instead, we should be loyal to ourselves and our communities. Money needs to come out of politics. This is why we must create a movement that solely relies on grassroots approaches and community-centered priorities.

Addressing systemic inequities inside and outside of the healthcare field is imperative in the next couple of generations. We should use the disciplines we have learned in this book to focus on the social and cultural factors related to power distributions and determinants of health in societies as an interdisciplinary perspective that combines several different fields. Some of these fields include epidemiology, ethnic-demographics, psychology, sociology, history, and medicine. By integrating these different approaches, we can move forward to understand and address problematic health disparities. The topics and ideas that we learned in first part of this book will help us systematically study the health and well-being of individuals who have been historically marginalized and develop programs and policies that can begin to repair much of the harm done to these communities. By learning and applying these ideas and concepts, the next generation of physicians will aid policymakers in developing laws that protect and safeguard health and well-being of all.

We Are the Future and the Challenges That Lie Ahead
★ ★ ★

New Americans are the future, and in our lifetime, we'll already face several issues that we need to talk about and address. One of these central issues is the dangerous impacts of climate change. Climate change is probably one of the greatest threats to our survival as a human species in a global context. Whether we are talking about severe weather pattern changes, floods, drought, and deadly storms, we will need novel solutions to mitigate the impacts and find ways to begin to reverse the damage we have done to our earth. As the future, we must make addressing climate change a top priority because of the many health implications it has, not only in our lifetimes but in the generations to come.

Another challenge that is not new and that continues to plague our society, as we have learned, is racism. Overcoming racial tensions and finding ways to continue to fight for equity are imperative for New Americans to address systemic health inequities. Instead of moving forward, lately, it feels like we have been moving backward in time. Race is so integrated into America that it will not be going anywhere anytime soon, or even ever. Therefore, we must find a way to reconcile racial differences and ensure that we move away from being a racist society. We, as the future, must understand that there is a difference between focus and exclusion. If something matters, this does not imply that nothing else does. The Black Lives Matter movement arose in the context of people receiving messages from a culture in which they live that their lives are less important than others, hence their call to action. When I say Black lives matter, it does not mean that other lives don't matter. So, it is a gross distortion of reality to scold them for not being inclusive enough when they say, "Black lives matter." Black lives mattering is about focus, not exclusion. Seeing the world and its people in mutually exclusive either-or terms will limit you. If you wish to bear the intellectual consequences of construction ideology, that's your decision, but know that it will impede your progress to become a good person.

We must always reframe our understanding and ask critical questions to achieve equity in American society. Black lives matter. Indigenous lives matter, Asian lives matter, European lives matter, but none can matter until they all equitably matter because all oppression is connected. All of our struggles are bound to each other, and much recognition is needed to develop a new vision. Uniting and collaborating right now is of most importance because time is running out before we further ruin the planet that the next generation will inherit.

I have made it a goal to build a movement that ensures that health is a priority in non-health sectors such as the transportation, education, economic, and criminal justice systems. I will continue to facilitate these types of discussions and connections in ensuring that we move toward a society that will eliminate health disparities and lead the world in health status and outcomes. To provide high-quality health care to all people, we need to start moving beyond coverage to create access to patient-centered care. It needs to be essential to meet patients and recognize that no one size fits all. We need to ensure that everyone will have high-quality care by making public health the corner of my medical practice. We need to continue to dismantle traditional barriers that prevent people from having health care by developing and advocating for policies while empowering my patients to be the leaders of their lives.

Conclusion

★ ★ ★

In this book, I have laid out the fundamental theories and practices of many topics that must be engaged to make robust decisions with the goal of achieving health equity. Learning the basics of public health and medicine is one of the cornerstones of overcoming health inequities. At the same time, learning the basics of critical race and ethnic studies is just s equally important and highlights the art of medicine. I provided different examples of the ways these concepts can play out in real life throughout the book. I have also given you a glimpse of my journey, putting you in my shoes as I made my way from the margins to medicine. Being able to mentally visualize several of the case scenarios I've described is essential to learning and developing skills in public health and ethnic studies.

As New Americans, we must be health equity advocates and continue to fight for the right for the entire human race to live a healthy life. I know that we have the potential to turn things around in our country and begin to heal the several hundred years of harm. Once health equity is achieved, health disparities will be eliminated. This book has shown that it takes passion and science to address the root causes of health disparities. Health equity is a science that addresses the social determinants of health and the implicit biases that people have been socialized to have.

The next steps we must take include performing rigorous, well-completed scientific research to identify evidence-based interventions. This means that we have to find out what works and what is backed by evidence to inform the decisions made in health care and our personal lives. In this book, I have laid out the concepts and theories that will allow us to advocate for universal access to health-care coverage and address basic life needs, including job security, access to high-quality and affordable education, and having safe neighborhoods. Overall, as New Americans, we must dedicate our lives to our communities and heal people when they are sick and combat our society's social ills.

Moving forward, when serving a diverse community, it is essential to continue researching and be knowledgeable of the community we are working with to identify the root causes of the issues at hand rather than making assumptions that we know what the community needs are. Serving the community's needs requires valuing all individuals within the community equally, recognizing and addressing historical injustices, and equitably providing them with resources. We can be more aware of the identities, biases, and privileges in our lives with adequate preparation and self-reflection. Being an ally is not a noun; it is a verb. With comprehensive intercultural knowledge of the community, it will be more likely that the people we serve will accept our allyship to address their true needs and concerns.

This book is a call for diverse solutions to address the concerns of the community while at the same time sending a message to those First-Generation students who aspire to become New Americans. At the basic, fundamental level, we know that we need to address health inequities in a holistic manner that integrates the psychosocial and physical well-being of people. The ideas discussed in this book are foundational to our society if we hope to move towards one that can eliminate health inequities. Addressing and incorporating all aspects of the social determinants of health will be the foundation to create a sustainable and healthy society for diverse future generations.

Acknowledgments

★ ★ ★

I want to acknowledge Kimberly Lee who was the professional editor I had the privilege to work with. Thank you for your thoughtful yet critical review of this manuscript.

I want to acknowledge the people who were generous enough to donate to the development and launching of this book.

I want to acknowledge the Self-Publishing School, in particular my coach and Chad.

I want to acknowledge Brenda J. Allen, Teresa de Herrera, Eric Wagner, and Brissa who all helped review my manuscript.

I want to acknowledge everyone who has supported me in my journey to become who I am today. It hasn't been easy for anyone, but here we are, dreaming. Life has given me several precious gifts, but the most significant was to be blessed with people who believe in me. Thank you family, thank you mentors, thank you friends. Thanks to everyone who in one way or another have done something to make me the person I am today. For you, who read the book, I invite you to open your mind and to dedicate your life to living your New American Dream.

Can You Help?

★ ★ ★

Thank You for Reading My Book!

I really appreciate all of your feedback, and
I love hearing what you have to say.

I need your input to make the next version of
this book and my future books better.

Please leave me an honest review on Amazon letting
me know what you thought of the book.

Thanks so much!

William Mundo